NEW COSMOLOGICAL MODEL
BASED ON THE THEORY OF SPECIAL RELATIVITY

ABSTRACT

After checking the shortcomings that force the Standard Cosmological Model to adopt non-scientific theories such as Dark Energy, Dark Matter and Cosmic Inflation; a new Cosmological model is planted, based on the principle of exchange between matter and energy established by the Special Relativity Theory. Through the equations that relate matter and energy to movement we were able to explain and at the same time demonstrate: the creation of the universe, the creation of space-time, the speed of expansion of the universe, the creation of primitive stars, the continuous process of creation of galaxies, the gravity, the stellar and primordial black holes, the intergalactic movement, the final destination of our universe with the creation of new Big Bangs and that we are part of an infinite Fractal Multiverse.

The General Theory of Relativity is discarded for its incompatibility with the expansion of the universe and Kepler's first law and also for its erroneous explanation of the precession of mercury's perihelion. Also discarded as unnecessary, in the New Cosmological Model of Special Relativity, the theories of Dark Energy, Dark Matter and Cosmic Inflation.

Finally, we contrast the evidence of the new theories resulting from the research with those corresponding to the Standard Cosmological Model, based on the theory of General Relativity.

First Edition / Update: 23/04/2023

INDEX

1 INTRODUCTION 7
2 THE CREATION OF THE UNIVERSE 11
   2.1 Reflections on Primal Matter 12
      2.1.1 The primal matter of the New Cosmological Model 12
      2.1.2 The primal matter of the Standard Cosmological Model 13
3 THE CREATION OF MATTER AND SPACE-TIME AND THE FORM AND DIMENSIONS OF THE UNIVERSE 16
   3.1 Creation of matter and space-time 18
4 The miniverses 26
   4.1 In the Cosmological Model Theory of Special Relativity Cosmic Inflation did not occur 27
   4.2 There are 56 miniverses 27
   4.3 In the Theory of the New Cosmological Model, the Theory of Dark Energy is unnecessary 28
5 THE SHAPE AND DIMENSIONS OF THE UNIVERSE ACCORDING TO ASTRONOMICAL OBSERVATIONS 28
   5.1 New concept of the Observable Universe 31
6 THE CREATION OF THE STARS AND GALAXIES OF THE UNIVERSE. 33
   6.1 Late Anisotropy confirms the theory of this research on the creation of primitive stars and galaxies. 35
7 TIME IS CREATED BY THE MOVEMENT OF MATTER 37
   7.1 Time does not dilate with speed 39
      7.1.1 "In an expanding universe, time is not distorted by movement at high speeds" 44
8 SCIENTIFIC EXPLANATION OF HUBBLE'S LAW AND PROOF OF THE EXPANSION MODEL OF THIS RESEARCH 44

- 8.1 The expansion of the universe is not accelerated 52
- 8.2 Clarification on the value of Hubble's parameter and its variation over time..53
- 8.3 Conclusions of the chapter of Hubble's law 54

9 THE CURVATURE OF SPACE-TIME, IN THE THEORY OF THE NEW COSMOLOGICAL MODEL 55

- 9.1 The spherical cap of the curvature of space time..........................57

10 GRAVITY IN THE NEW COSMOLOGICAL MODEL 58

- 10.1 The cause of gravity and the meaning of the gravitational constant G   58
- 10.2 The gravitational field generated by a spherical body....................61
- 10.3 Gravity between two spherical bodies..62
- 10.4 Gravity in spheroidal bodies...65
- 10.5 Gravity between two spheroidal bodies 67
  - 10.5.1 Analysis of the results of the gravity experiment between two spheroidal bodies 77
  - 10.5.2 Calculation of the orbit of the planet Saturn around the Sun, based only on the data of its geometry and the inclination of its axis of rotation. 79
  - 10.5.3 Theory of "The Cause of the Inclination of the Axis of Rotation of Earth and Other Planets" 82
  - 10.5.4 Theory of "The cause of the elliptical form of planetary orbits defined in Kepler's first law" 82
- 10.6 Time is not changed by the effect of gravity 83
- 10.7 Conclusions on the origin of gravity in the New Cosmological Model   83

10.8 The model of the curvature of space-time of the Theory of the New Cosmological Model is confirmed 86

10.9 Theory of modification of the law of universal gravitation............87

10.10 Einstein's theory of General Relativity is discarded, as it is incompatible with the expansion of the universe and Kepler's First Law. ..88

10.11 Light is not curved by the effect of gravity 88

10.11.1 Theory: "Light is observed by describing a curved trajectory in the universe" 89

10.11.2 The explanation of the curvature of light observed in Arthur Eddington's experiment in 1919   91

10.11.3 The cause of the phenomenon called Gravitational Lenses 92

10.12 There is no anomaly in the mercury orbit 92

10.12.1 The angular acceleration envisaged in Kepler's 2nd Law solves the apparent anomaly of Mercury's orbit 95

10.13 The Sun's hidden orbits..96

10.13.1 The spinning elliptical Orbits of the Sun   98

10.13.2 The cause of perihelion precession. ... 113

10.13.3 Astronomical observations confirm the Theory of the Sun's spinning Orbits........... 113

11 THE BLACK HOLES.................................... 116

12 STELLAR BLACK HOLE 117

12.1 Radial flow of space-time..119

12.2 Gravitational energy of a stellar mass black hole 120

12.3 Dimensions of the components of a stellar black hole 126

12.4 The flow of space-time into a stellar black hole, called kerr flow. 128

12.5　The acceleration of gravity of a stellar mass black hole 128

12.6　The phenomenon of pulverizing the attracted mass (tidal disruption event) 128

12.7　Process of absorption of matter captured by the black hole 129

13　INTERMEDIATE-MASS BLACK HOLES 129

13.1　"The drag force of space-time" 129

13.2　Process of joining two black holes 130

14　SUPERMASSIVE BLACK HOLE 130

14.1　Basic information about galaxies 130

14.2　Analysis of the functioning of a galaxy based on the theory of stellar black holes 132

14.3　Mass of the black hole of the galaxy NGC 4258 133

14.4　Velocity of stars on the periphery of galaxies 134

14.5　The Dark Theory of Dark Matter. 134

14.6　The Theory of Stellar Black Holes of the New Cosmological Model Explains the Motion of Stars on the Periphery and Inside galaxies 136

14.7　Galactic bulb 138

14.8　Process of transferring stellar flow from galactic arms to the disk of the galactic nucleus 138

14.9　The Bar of Barred Spiral Galaxies. 143

14.10　Illuminated area of the nucleus of elliptical galaxies. 145

14.11　Final stretch of stellar flow 146

14.12　Process of absorption of stellar flow by SMBH 147

14.13　Dimensions of the SMBH of the galaxy NGC 4258 147

15　PRIMORDIAL BLACK HOLES 148

15.1　The Drag flow of space-time into the interior of primordial black holes. 153

15.2 The gravitational energy of a primordial black hole. 153

16 THE QUASARS 155

16.1 Theory: The origin of quasars 157

16.2 Astronomical observations confirm the theory of the origin of quasars 158

17 THE INTERGALACTIC SYSTEM 159

17.1 Theory: Primordial Flux determines the motion of galaxies 160

17.2 Motion simulation of galactic groups 161

17.3 The primordial flux is not altered by the acceleration of gravity due to the gravitational energy of the primordial black hole. 163

17.4 System of recycling of the matter of the universe 163

17.5 Evaluation of the experiment of simulation of the movement of galactic groups 163

18 THE END OF THE UNIVERSE 166

18.1 From the Miniverses to the Fractal Multiverse 166

19 THE ORIGIN OF THE BIG-BANG 168

20 CONCLUSIONS 168

20.1 Analysis of the evidence of proof of the New Cosmological Model 170

20.2 Analysis of the evidence of the Standard Cosmological Model Theory (ECM) 176

21 ACKNOWLEDGEMENTS 181

22 REFERENCES 182

23 AUTHOR DATA 183

1 INTRODUCTION

The vast majority of universities and research centers of astrophysics and astronomy have adopted the Standard Cosmological Model as the fundamental basis for the teaching and research of science that studies the universe.

The Standard Cosmological Model has as its scientific basis the Theory of General Relativity created by Albert Einstein in 1915, which emerged from his Theory of Special Relativity, designed for an inertial system, as an adaptation of that theory to a gravity-accelerated system. However, the universe, better known in our time, is not an accelerated system; it is clearly an inertial system. This has been demonstrated by space travel, because anywhere in space away from a massive body, a small impulse on an object is enough for this object to initiate unstoppable movement at constant speed. We could define the universe as an inertial system with small islands of massive bodies, where in its vicinity there is an accelerated field of attraction. So if we navigate into space we necessarily do so under the laws of inertial states until we reach the vicinity of a massive body, where we can take the impulse of the gravitational acceleration of the massive body and inertially continue the trajectory. And it is precisely the process described above that is used in space travel. Whose best example are the Voyager I and II spacecraft, which have more than 30 years traveling inertially without disturbances, one of them being already outside the solar system. On the other hand gravity decreases with the square of the distance, so its influence is reduced to bodies that are very close, such as planetary systems and is not applicable between bodies that are at great distances such as the stars within galaxies and the intergalactic movement. From the arguments set out above we conclude that the Theory of General Relativity is an adaptation of the Theory of Special Relativity to a gravity-accelerated system that does not exist in our universe. On the other hand, the Standard Cosmological Model considers the great explosion of the Big-Bang as the origin of the universe, it seems of elementary logic that our universe must be made up of the residues of that explosion therefore its matter and energy must come from it. However, the Standard Cosmological Model states that after the Big-Bang there is no residue of the explosion and matter is created from elementary subatomic particles, adopting a creationist model similar to that of theological theories, without explaining where the immense energy is obtained for that creation. In addition, a static primary matter does not create the movement necessary for the creation of time, since time is a consequence of movement and if there is no movement there is also no time. It is a scene of a universe failed for lack of energy and time for its development and evolution.

The discrepancy between the theory of General Relativity, which presupposes the existence of a gravity-accelerated system, with the inertial reality of the universe and the shortcomings of the hypothesis about its primeval matter, makes the Standard Cosmological Model unworkable to rightly study our universe. As a result of this incompetence the Standard Cosmological Model resorts and assigns to three theories without scientific basis a relevant role in explaining the origin, operation and destiny of the universe: Dark Energy, Dark Matter and Cosmic Inflation. Dark Energy and Dark Matter have that name "dark" because of their unknown nature. However, the Standard Cosmological Model states that Dark Energy brings 70% of the mass-energy of the universe, dark matter contributes 25% and I only corresponded 5% to ordinary matter. Which means, according to this theory, humanity since the cave season, with all the technological development it has achieved in its history, has only advanced 5% in understanding the universe.

The formulation of these three theories is contrary to the scientific method which states that scientific theories are formulated to explain the observed reality.

In the case of dark matter because gravity is not competent, by the great distances involved, to explain the movement of stars within galaxies and intergalactic movement, the existence of invisible matter contrary to observed reality is assumed to satisfy compliance with the theory, which is clearly contradictory to the scientific method.

In the case of Dark Energy and Cosmic Inflation, both theories have been formulated only to justify the Theory of the Standard Cosmological Model on Primal Matter which is a static and energyless matter, which is contradictory to its origin of a highly energetic event such as the Big-Bang and incompatible with the expansion of the universe, widely demonstrated reality.

Belief about these three theories without scientific basis represents a "headache" for the world of science, which represents a stumbling block for study and the best understanding of the universe, as they act as limiting ideas that have prevented, to this day, knowledge of basic and elementary things in the universe, such as:

1) The current shape and dimensions of the universe
2) The speed of expansion of the universe
3) The source of the energy of the universe
4) The Exact Age of the Universe
5) The Cause of Gravity and exact value and what represents the gravitational constant G
6) The Scientific Explanation of Hubble's Law
7) The Scientific Explanation of Kepler's First Law
8) The Origin of the Big-Bang The Fate of the Universe

Studying the universe is akin to crossing a complex maze full of parallel paths, many of which lead us to dead end roads, and that seems to us to be happening with the Standard Cosmological Model, we find ourselves at the end of a blind street. What can be done in a case like this? The usual thing in those cases is to return and look for a parallel path that can lead us to the exit. And that's exactly what we're going to do in this investigation. We will return to the Big-Bang, to the starting point when the universe was created and from there we will develop a new theory alternates to the Standard Cosmological Model.

For the development of this new theory, which we will call "New Cosmological Model", we will use as a scientific basis the Theory of Special Relativity created by Albert Einstein in 1905 for the study of inertial systems, in full concordance with the inertial reality of the universe. The fundamental basis of the Special Theory of Relativity is the principle of equivalence and interchangeability between energy and matter. It is a notorious fact in nature the conversion that exists between energy and matter. It is an exchange that flows like an eternal symphony, where time marks movements. It is present in all natural phenomena, especially in life. Living things are not created out of nowhere, are born, grow and die due to the transformation of energy into matter and vice versa.

The same goes for the universe or universes. They were never created out of nowhere. They are born grow and die due to energy-matter-energy transformation in a cycle that is permanently repeated.

The convertibility between energy and matter and its relationship to movement, allows us to explain and scientifically verify, the whole process of creation, evolution and destiny of the universe through the following equations, developed by Albert Einstein, in his Special Theory of Relativity:

$$E = \sqrt{p^2c^2 + m_0^2c^4} \quad (E1-1)$$

$$m = \frac{m_o}{\sqrt{1 - \frac{v^2}{c^2}}} \quad (E1-2)$$

Equation E1-1 measures energy growth with movement and equation E1-2 measures the growth of matter with motion.

According to equation (E1-2), the mass of a body moving at the speed of light grows to infinite values. But for this to happen, infinite time has to elapse.

When you enter the time variable, equation (E1-3) is derived from equation (E1-2):

(E1-3) $\quad x = (dx + c\,t) \;;\; y = (dy + c\,t) \;;\; z = (dz + c\,t); Z = (dZ + ct)$

For a body that moves at the speed of light, the first three terms of the equation measure the growth in body time at its three coordinates (x,y,z) and the fourth term, measures the movement of the body on the Z axis (hyperspace), parallel to the axis of time.

These three equations, allows us to scientifically demonstrate the following:

1) The creation of the universe.
2) The shape and dimensions of the universe.
3) The expansion of the universe.
4) The creation of matter and space-time.
5) The creation of primitive stars and galaxies.
6) The movement and speed of bodies in space-time.
7) The curvature of space-time and gravity.
8) The shape and operation of black holes and their potential energy.
9) The movement of stars within galaxies.
10) The intergalactic movement.
11) The energy that will create the new Big-Bangs and consequently, by analogy, the origin of the Big-Bang, and the conclusion that we are part of an infinite Fractal Multiverse in space and time.

Thus the Theory of the "New Cosmological Model", based on the convertibility between energy and matter offers a scientific explanation of the whole universe.

2 THE CREATION OF THE UNIVERSE

One consequence of movement is time. There is no time if there is no movement and no movement if matter does not exist. Therefore matter could not be created out of nowhere, because without matter there is no movement and therefore there would be no time for its creation.

The above reasoning means that our universe was not created out of nowhere, it emerged as a change in the state of energy and matter, so we can explain it using equations (E1-1), (E1-2) y (E1-3)

Before the Big-Bang, a singularity probably belonging to a black hole from an earlier universe, reaches an almost infinite value of its density and temperature. In the case of no movement, the equation of energy (E-1) is reduced to the second term i.e. that corresponding to the mass at rest:

$$E=mc^2$$

Matter has to have limits of compression, density and temperature, exceeded that limit there was an immense explosion and as a result of it, the singularity of fracturing in "n" parts, which shot out in all diver-gent directions at an angle "θ", until the speed of light is reached. The finite number of "n" parts in which the singularity was fractured, de-spite the magnitude of the explosion, is explained by the almost infinite density of it.

The "n" particles of extreme density due to movement increase their mass, as established by the equation (E1-3), reaching within a period of its path to the limit of the critical mass of nuclear fission, resulting in a chain reaction of nuclear fissions. The chain of fissions culminates in reaching the hydrogen, the lightest material that exists. In this way at the end of the process we have the primal matter of the universe, which will consist of "n" infinitesimal particles of hydrogen moving at the speed of light.

The paths traveled from the site of the explosion by the "n" particles in all directions, during the cycle of nuclear fissions, form a sphere, by joining the particles together, and is what we now know as the CMB Cosmic microwave Background.

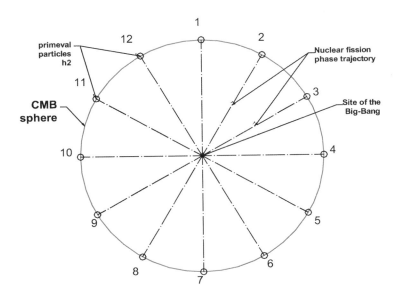

Figure 1 Primeval particles arising from the Big-Bang

From that moment our universe begins and time is created as we know it associated with the movement of primal particles at the constant speed of light. Therefore, "n" axes of divergent times were created according to the angle "θ" and synchronized so that each of them marks the same time since the creation of the universe

We have found that our universe was not created out of nowhere, it emerged as a change in the condition of matter and energy, by which a mass at rest of maximum density was transformed into "n" masses of minimum density moving at the maximum speed of light

2.1 Reflections on Primal Matter

The hypothesis of primal matter is a turning point between the theory of the Standard Cosmological Model and the theory of the New Cosmological Model. Next we will evaluate their differences and the evolutionary path to which both hypotheses lead us:

2.1.1 The primal matter of the New Cosmological Model

According to the New Cosmological Model, primal matter is "n" particles that move in divergent trajectories according to an angle "θ" at the speed of light in the field of an inertial system. The following consequences follow::

- The motion of the primordial particles in the inertial system of the universe is permanent at the speed of light, which results in more energy also being permanently created according to the equation. (E1-1).

- The dimensions of matter in motion grows permanently at the speed of light, according to the equations (E1-2) y (E1-3)
- Evolution occurs identically in the "n" directions according to the angle "θ", which guarantees an evolution of the isotropic and isothermal universe in all directions. This is consistent with the observed reality.
- Motion in all directions explains the expansion of the universe from both space and matter at the speed of light.
- The speed of light does not determine for itself the maximum possible speed in the universe. It is an indicator of that maximum speed but does not determine it, as a thermometer cannot determine the temperature of a body. The maximum speed of the universe is determined by the speed at which the primal particles are moving away, since nothing can exceed the speed of expansion of the universe.
- Movement in "n" divergent trajectories creates synchronized "n" time axes and a growing spherical shape of the universe at the speed of light.
- All the growing energy of the universe originates from the remaining energy of the great explosion called Big-Bang inherited by its primal matter.

2.1.2 The primal matter of the Standard Cosmological Model

According to the Standard Cosmological Model after the Big-Bang occurs, there is no constituted matter, matter is caused by an evolution of subatomic particles that are first transformed into atoms and then hydrogen molecules, later evolving with the creation of helium, then hydrogen and helium multiply and form molecular clouds of hydrogen and helium. Molecular clouds by acretion processes would then give rise to stars and galaxies. The following conclusions follow from this primal matter hypothesis:

- The creative process disassociates the primal matter from the energy of the Big-Bang
- There is no explanation for the physical process by which a matter, of almost infinite density and temperature, after an explosion is reduced to simple subatomic particles. The physical process by which a high-density matter part is transformed into a lower density is known as nuclear fission and a chain of nuclear physions irretrievably ends up in the lowest density element that is molecular hydrogen.

- There is no explanation for where the energy that allows atoms to form and later hydrogen and helium molecules and even less so in the large amounts of molecular clouds that would form stars and galaxies.
- In a flat universe, static primary matter without energy inherited from the Big-Bag cannot generate the expansion of the universe. Faced with this deficiency, the Standard Cosmological Model formulates a theory whose name Dark Energy means is that it is not known where that energy comes from.
- In a flat universe static primal matter cannot explain the isotropic and isothermal evolution of the universe, so the Theory of the Standard Cosmological Model adopts the theory of Cosmic Inflation, which may explain this problem but creates another of equal magnitude, what is the source of energy to produce the Cosmic Inflation?

By evaluating both theories on primal matter we can conclude:

The theory of primal matter formulated by the Theory of the Standard Cosmological Model is unworkable as it is incompatible with events that happen in the universe properly proven, such as the expansion of the universe and its isotropic and isothermal condition in all directions and by its lack of energy and axes of time for its evolution and development.

On the contrary, the theory of the primal matter of the New Cosmological Model is not only compatible with the expansion of the universe and its status as isotropic and isothermal in all directions but rather is the very cause of both phenomena and gives the universe energy and time for its evolution and development.

But the determining evidence in favor of the theory of primal matter of the New Cosmological Model is the evidence found in the Microwave Emission Fund, which reveals that the universe has a growing spherical shape as a balloon that is permanently inflated in perfect coincidence with the theory of the New Cosmological Model. An article published in Nature Astronomy[1] has found an anomaly in the CMB that can only be explained by a closed universe.

Let's look at a summary of this article published in the ABC newspaper[2] of Spain. Signed by José Manuel Nieves of 11/05/2019

---

[1] Planck evidence for a closed Universe and a possible crisis for cosmology | Nature Astronomy
[2] ¿Vivimos en un Universo en forma de bucle? (abc.es)

"Everything we thought we knew about the shape of the Universe could be wrong. According to a new study recently published in Nature Astronomy by researchers from the universities of Manchester (United Kingdom), La Sapienza (Italy) and John Hopkins (USA), the Universe could be curved, like a huge balloon that swells non-stop, rather than being flat as a sheet, as current theories postulate.

Researchers have come to that conclusion after finding an "anomaly" in the cosmic microwave background (CMB), the weak echo of the Big Bang that permeates the entire Universe.

If the Universe were curved, as the study points out, it would mean that, just as it happens on Earth, we could move in a straight line and end up, at some point, returning to the starting point. It's what's known as a "Closed Universe."

Of course, curvature occurs on such a large scale that it is difficult to perceive it locally, from Earth, the Solar System or even at the galactic level. It would take a much broader perspective to realize.

The point is that for decades now cosmologists dismissed the idea of a closed Universe, because it does not fit at all with existing theories about how Cosmos works. Instead, the idea of a "Flat Universe" has been imposed that extends without limit in all directions and that at no time folds over itself.

But now, in the best measurement made so far of the CMB, an anomaly in that background radiation seems to indicate that, after all, we could be living in a closed Universe. According to the current inflation model, the Universe should be open. . In the first billionths of a second after the Big Bang, the model says, there was a moment of exponential expansion during which the Universe went from being a simple point to a specific physical space. And the physics of that super-fast expansion point to a flat universe. That is precisely why most physicist opts for this option today. But if the Universe turned out not to be flat, you would have to "adjust" all physics to that new reality, and perform a huge number of other calculations. Something that, according to the authors of the work, might be needed very soon.

An anomaly in the CMB

What is the anomaly detected in the cosmic microwave background? The CMB is the oldest element we can see in the entire Universe. It consists of a faint "environmental" microwave radiation that floods all space and is one of the most important data sources on the history and behavior of the

Universe as a whole. Well, according to the data from the last measurement, the most accurate to date, there are a much larger number of CMB "gravitational lenses" than would be expected, and that means that gravity could be "bending" the CMB microwaves much more than current physics is able to explain.

The researchers themselves, however, indicate that, although the evidence is solid, their results are not entirely conclusive. According to the calculations carried out by the team, the data point to a closed Universe with a standard deviation of 3.5 sigma (a statistical measurement that means that there is a 99.8% chance that the result is not caused by a statistical error). And that's still well below the 5 sigma that physicist needs before confirming an idea.

The debate, then, is served. The study has found a significant discrepancy, an anomaly that needs an explanation. And the one in the closed Universe could be the most valid. However, new research could bring other ideas that help solve the problem. For now, suffice it to say that it is possible that the idea that we live in a closed Universe has been discarded too quickly".

The permanently growing spherically closed universe is one of the most important conclusions of the New Cosmological Model, and arises precisely from the theory of primal matter set out in this chapter. In the next chapters this idea will be expanded with the precise measurements and growth rate of the spherical form of the universe.

The following theories originate from the theory of primal matter of the New Cosmological Model:
- The Creation of Matter and Space-Time
- Theory of Miniverses Cosmic Inflation is Unnecessary in the Cosmological Model of Special Relativity
- Dark Energy is Unnecessary in the Cosmological Model of Special Relativity

These theories will be described in detail in the next chapters.

3 THE CREATION OF MATTER AND SPACE-TIME AND THE FORM AND DIMENSIONS OF THE UNIVERSE

We observe again Illustration 1, the "n" primordial particles of hydrogen begin their trajectory from the periphery of the CMB, starting the time of the universe.

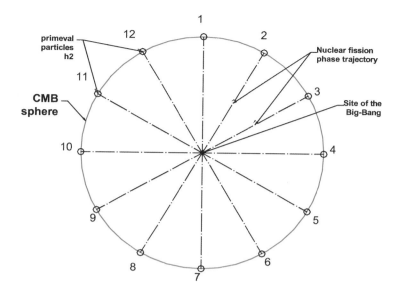

Figure 1 Star of the universe
The material bodies, arising from the explosion of the Big Bang, begin the second phase by continuing their trajectory from the periphery of the CMB at the speed of light, they are "n" divergent particles at an angle Θ of unknown value

A body moving at the speed of light permanently increases its energy and therefore its mass according to equations (E1-1), (E1-2) and (E1-3). If we apply equation (E1-3) to analyze the trajectory and growth of primeval particles due to their movement at the speed of light, the result would be that they will travel a path equal to the radius of the observable universe r0, and will grow in the three coordinates a value equal to the radius of the observable universe r0. Consequently the spheres of the primordial particles will become tangent when the separation between them is equal to the radius of the observable universe. In Figure 2 we can observe the universe as the spheres derived from the displacement and growth of the primordial particles become tangent. Since the distance traveled by all particles is equal to the radius of the observable universe r0, and the distance between the centers of the contiguous spheres is also equal to r0, it follows that in all cases an isosceles triangle is formed with an angle of 30° at the center of the universe. This means that the angle "Θ" between the trajectories of the particles is necessarily 30° and therefore there are 12 particles in the plane and 56 in the sphere of the universe as we will demonstrate later.

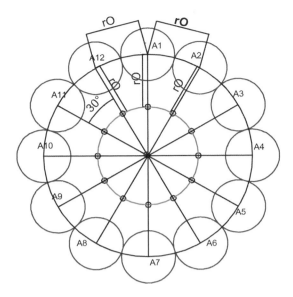

Figure 2 Hydrogen spheres

As a result the universe would be constituted by 56 immense tangent spheres of hydrogen of radius 0.5rO, without the existence of space-time or the bodies that are housed in it, stars, galaxies, etc. But we know that didn't happen..

But then how did the 56 particles of primordial matter travel the path of the radius of the observable universe rO, and transform into the immense amount of matter found in the almost infinite number of galaxies that exist in the universe?.

3.1 Creation of matter and space-time

To solve this paradox we will use the equation (E1-1), for a body that moves at the speed of light.

$$E = \sqrt{p^2c^2 + m_0^2c^4} \quad (E1-1)$$

$$E = (pc \rightarrow) + (m_o c^2 \rightarrow)$$

P= mv. In this case v=c therefore P=mc y $m=m_0=m_p$; m_p beig the primal matter. Replacing:

$$E = (m_p c^2 \rightarrow) + (m_p c^2 \rightarrow)$$

The two terms have the velocity vector moving in the same direction, therefore they are two parallel vectors of equal direction and sense. Which is why we add its modules algebrially:

$$E = m_p c^2 + m_p c^2 = 2\ m_p c^2$$

E= 2 $m_p c^2$ (E3-1)

Therefore we have an energy increment vector following the trajectory of each of the 56 particles of primeval matter.

A body moving at the speed of light will be able to double its energy and consequently its mass, in accordance with the principle of the Theory of Special Relativity which states that "the energy released is equal to the disintegrated mass multiplied by c² and the energy absorbed is equal to the created mass multiplied by c²"

This means that a body traveling at the speed of light has two options: 1) it grows at the speed of light; 2) does not grow and doubles creating identical clones.

This is a novel revelation, which allows to solve the paradox as follows: The primordial matter, which travels at the speed of light, before growing on its axis of time, first doubles, in geometric progression, creating identical clones in the contiguous time axes, creating concentric rings where the clones of the primordial matter are deposited with a number of twice the preceding matter. The process is repeated creating new and larger circles containing cloned primeval matter. That is, matter and space-time are created simultaneously where it will be located.

For this to be possible using the principle of equivalence the time axes where the clones will be located, must correspond to the growth of a spherical body that has twice the mass and therefore twice the volume to the cloned body, therefore:

$$V_2 = 2V_1$$

$$\frac{4\pi}{3} R_2^3 = 2 \frac{4\pi}{3} R_1^3$$

$$R_2 = \sqrt[3]{2}\, R_1$$

$$R_1 = x/2$$

$$R_2 = (x + ct)/2$$

$$\frac{x + ct}{2} = \sqrt[3]{2}\, \frac{x}{2}$$

$$ct = 0{,}26\, x$$

As a consequence the location of the clones will be in circles of radius $\sqrt[3]{2}$ greater than the previous one starting from the diameter of the primeval particle

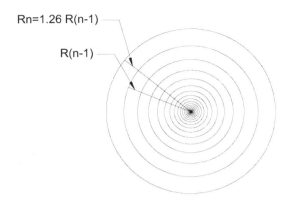

Figure 3 Concentric circles of location of clones of primeval matter

At the beginning of the movement we have a unit of the primeval matter and at the end the sum of clones of the primeval matter equivalent to the mass of the final sphere that would result from a growth according to the equation (E1-3). That is:

$$ME(0.5r0) = \sum_{n=1}^{x} 2^n \, m_p$$

$$\frac{4\pi}{3}(0.5r0)^3 = \sum_{n=1}^{x} 2^n \, m_p \qquad (E3-2)$$

The number of clones will depend on the size of the primeval particle. As a consequence of the massive cloning process, the hydrogen spheres that would be formed with the direct application of the equation (E1-3) are transformed into fictitious spheres whose function is to intercept the time axes where the clones were deposited (Figure 4).

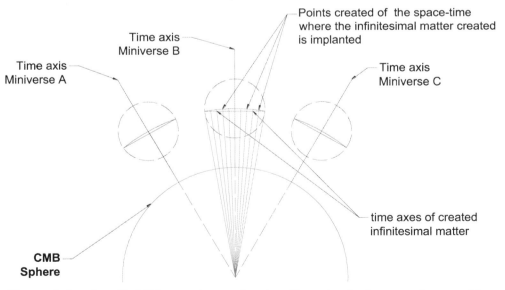

Figure 4 Growing fictitious spheres forming the spherical caps of space-time

This process continues until the spheres and spherical caps inside become tangent (figures 7 and 8).

In Figure 5 and Table 1 we can see in more detail a simulation of this process.

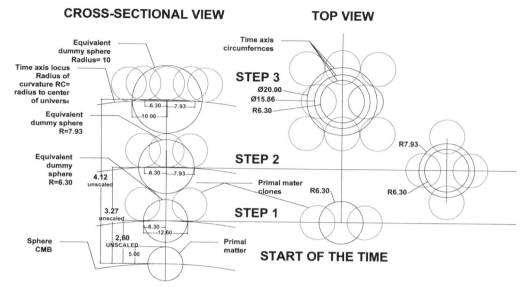

Figure 5 Process of cloning primordial matter and formation of space-time

Table 1. Formation spherical caps of space-time

STAGES	Primal matter diameter	Displacement	Equivalent sphere diameter	Radius circles e-t	Equivalent sphere diameter	Number of equivalents clones	Last circunference clones	Total sum of clones
START OF THE TIME	10,00	0,00				1		523,60
STEP 1	10,00	2,60	12,60	6,30	1.047,20	2	1	1.047,20
STEP 2	10,00	3,27	15,87	7,94	2.094,40	4	2	2.094,40
STEP 3	10,00	4,12	20,00	10,00	4.188,80	8	4	4.188,80
FINAL STEP	10,00	$r0$	$r0$	$0,5r0$	$4\pi/3(0.5r0)^3$	2^n	2^{n-1}	523.6×2^n

1) Start of the time: Each of the 56 primeval particles of diameter 10 (symbolic units US) at the end of the fission process, is located on the outer axis of the surface of the CMB sphere, and starts driven by the remaining energy of the Big-Bang, its trajectory at the speed of light until it reaches the radius of the observable universe **r0**.

2) **Step 1:** When traveling a distance 2.6 US according to equation (E1-3) they are transformed into a sphere of diameter 12.6 US, and according to equation E1-1 doubles their energy, which is equivalent to two identical bodies (clones) located on the contiguous time axes in a circle of diameter 12.6 US

3) **Step 2:** When reaching the path a distance of 3.27 US according to equation (E1-3) is transformed into a sphere of diameter 15.87 US, and according to equation (E1-1) again doubles its energy, being equivalent to 4 clones. Therefore to the two clones arising from step 1 located in the circle of diameter 12.6 US, 2 clones are added in a concentric circle to the previous one of diameter 15.87 US.

4) **Step 3:** When reaching the path a distance of 4,12 US according to equation (E1-3) is transformed into a sphere of diameter 20 US, and according to equation (E1-1) doubles its energy again, being equivalent to 8 clones. Therefore to the 4 clones arising from the two previous steps are added 4 clones in a concentric circle to the previous ones of diameter 20 US.

The process will continue until the displacement reaches the distance of the observable universe r0, with the values indicated in the Table 1

In this way space-time is created, which we should call, space-time-matter, in the form of spherical caps whose radius of curvature will be equal to the length of the time axes. The space-time network as shown in Figure 6 is not square, it arises from the intersection of concentric circles of time and radial lines of space that joins the clones.

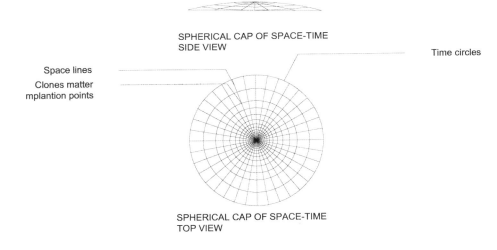

Figure 6 Spherical cap of space-time

At the points of intersection, as we explained before, are the time axes where the clones of primeval matter are deposited. As we can see, space-time is not an imaginary surface, it is something very concrete physical, where matter is housed, grows and can be moved. Something like an intergalactic platform or mothership, traveling and expanding at the speed of light, in the form of an umbrella, where the supporting structure of the umbrella is the axes of time and the fabric the surface of space-time. At the end of the journey the spherical caps finally become tangent and form the Surface of space-time, similar to the outer part of a globe. The radius of the Surface of space-time is equal to the sum of the radius of the observable universe rO, plus the radius of the CMB. (figure 7 and 8).

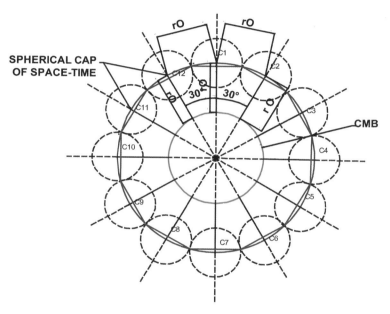

Figure 7 The universe and its spherical space-time caps. Cross-sectional view

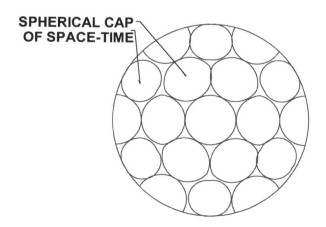

Figure 8 The universe and its spherical caps of the E-T. 3D Vision

Until that moment the spherical caps, which we will also call "Miniverses" were continuously separated from the sphere of the CMB, at the speed of light and grew at the speed of light, without any inconvenience. But when tangents become tangents, the following problems arise:

1) A jam occurs and the caps are prevented from growing and creating new circles of space-time.
2) Enormous pressure is created between the spherical caps as each cap is pressed by the 6 spherical caps that surround them, whose pressure force is concentrated in the center of each cap.
3) The mass of the center of each cap where the density is greater, collapses due to immense pressure and forms a primordial black hole.
4) As a consequence of the creation of a primordial black hole at the center of each Miniverse, a flow of the space-time mesh occurs to continuously fill the inner surface of the pit where the black hole is housed. The flow of the space-time mesh continuously releases the outer edge of the miniverse, being able to continue indefinitely the process of creating new points of space-time and implantation of the cloned primeval matter.

The spherical caps have to continue their movement which forces them to continue their growth, but both things are impeded by the jam. How is this conflict resolved? The solution is as follows:

The "Metric Expansion of Space begins"[3]. The metric that formally describes the expansion in the standard Big Bang model is known as the Friedman-Lemaître-Robertson-Walker metric, following which the universe expands as a single set. A continuous change of the metric and therefore of scale at the speed of light begins, which allows the caps to continue their trajectory because it increases the scale of the distance that separates them from the CMB, also expanding the sphere of the CMB. The distance will always be equal to the radius of the observable universe $r0$, but as $r0$ grows at the speed of light the platforms move away from the CMB at the speed of light without the relative distance growing, but the actual distance. The same happens with the dimensions of the caps, they fix their internal radius at a value of $0.5\ r0$, which also allows their growth without compressing

---

[3] https://es.wikipedia.org/wiki/Expansi%C3%B3n_m%C3%A9trica_del_espacio#:~:text=La%20expansi%C3%B3n%20m%C3%A9trica%20del%20espacio,hace%20m%C3%A1s%20joven%20o%20viejo.

against each other, keeping the distance between the centers of the spherical caps equal rO. Previously we had calculated the angle of 30° of separation of the spherical caps. In order to complete the design we must calculate the radius of the internal circumference, occupied by the Cosmic Microwave Bckground (CMB) whose center is the center of the Universe, which is obtained directly by measuring the distance from the center at the end of one of the lines that start from C1, C2 and C12.

It is also possible to analytically obtain the value of the radius of the inner circumference as follows:

The isosceles triangle C12, C1, O, has its 30° vertex at the center of the universe. The distance between the two vertices of 75° is the distance between A12 and A1, which is rO, as is already known. The other side has by length the sum of rO plus X, being X the distance that interests us, that is, the radius of the internal circumference. With this data, we will then calculate its exact value, using the trigonometric functions of the 15° angle:

$$Sin(15º) = \frac{1}{2}rO/(rO + x)$$

$$Sin(15º) = \frac{1}{2\sqrt{2 + \sqrt{3}}}$$

It follows accordingly: $x = rO\left(\sqrt{2 + \sqrt{3}} - 1\right)$

$$x = 0{,}93185\, rO$$

Radius of the sphere of the CMB, Rcmb= 0,93185 *rO*

In this way the model of the universe shown in illustrations 9 and 10 was created.

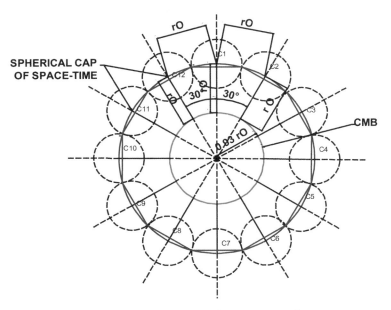

Figure 9 The universe and its spherical space-time caps. Cross-sectional view

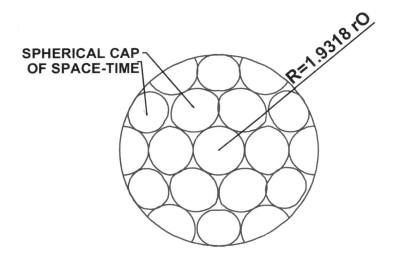

Figure 10 The universe and its 56 spherical E-T caps. 3D Vision

In this way (figure 10) we can affirm that the outer sphere of the universe where space-time and matter reside is not a continuous sphere but rather a discontinuous surface constituted by the 56 tangent spherical caps derived from the independent evolution of the 56 primordial particles

4   The Miniverses

The aforementioned tangent spherical caps come to be like a kind of small autonomous universes and, therefore, it is valid to affirm that the evolution of the universe was carried out separately, but at the same time in a synchronized way, in each of these 56 spherical caps. This results in the universe being homogeneous in all directions. It is also isotropic and preserves a thermal equilibrium at great distances, because the same thing, and at the same time, happened in the 56 spherical caps distributed symmetrically throughout the universe. The aforementioned elements act as small independent and contiguous universes, so we will call those spherical caps Miniverses..

Miniverses are autonomous and independent of each other and live in parallel all stages of evolution of the universe, from its birth to its end. Miniverses derived from primeval particles permanently increase their energy due to their perpetual motion at the speed of light.

As a consequence we formulate the following theory:

4.1 In the Cosmological Model Theory of Special Relativity Cosmic Inflation did not occur

The theory of the Miniverses explains that the universe is isotropic and isothermal in all directions; therefore in the Cosmological Model Theory of Special Relativity there is no need for a theory without scientific basis such as Cosmic Inflation whose objective is to explain that the universe is isotropic and isothermal in all directions.

4.2 There are 56 miniverses

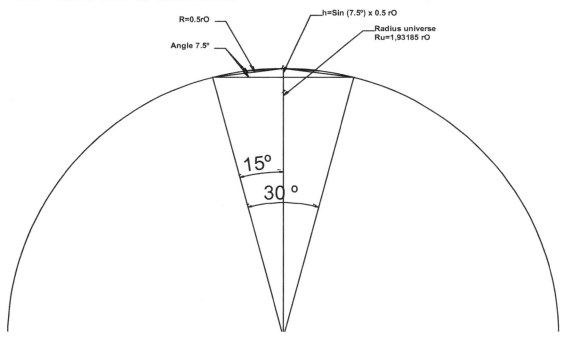

Figure 11 Spherical cap of a Miniverse

The area of the outer crust of the universe is:

$A_e = 4\pi \; R_u^2 = 4\pi \times (1.93185 r_O)^2$: $A_e = 46.898 \; r_O^2$

Spherical cap area: Ac= 2π hxRu=2πsen7.5°x0,5rO= 0,792 rO²

Empty space area per spherical cap (measurement by AutoCAD) ≅ 0,0451 rO²

Unit area= Au=0,792 rO²+0,0451rO²= Au= 0,8372rO²

Number of units: Nu= Ae/Au=46,898rO²/0,8372rO²= 56,018

Therefore: **Nu=56**.

Therefore it is shown that the universe expands at the speed of light both matter and space due to the energy of its 56 primordial particles now transformed into Miniverses, according to the equation $E = \sqrt{p^2c^2 + m^4c^4}$. As a consequence, we affirm:

4.3 In the Theory of the New Cosmological Model, the Theory of Dark Energy is unnecessary

The Dark Energy Theory[4] arose due to ignorance of the energy needed for the expansion of the universe. As we have seen that the universe expands at the speed of light due to the energy of its 56 primordial particles according to the equation $E = \sqrt{p^2c^2 + m^2c^4}$,, the theory of dark energy is discarded as unnecessary.

5 THE SHAPE AND DIMENSIONS OF THE UNIVERSE ACCORDING TO ASTRONOMICAL OBSERVATIONS

The shape and dimension of the universe calculated in the previous chapter can be confirmed by astronomical observations as we will demonstrate below:

The farthest we can see with a telescope represents a length equal to that traveled by light since its appearance in the universe, a distance at which we have defined as the "radius of the observable universe" (rO). And it turns out that the image that can be observed in all directions at that distance, is that of the Cosmic Microwave Background (CMB). Figure 12 shows what we have just mentioned, in two dimensions. Our observation site, point A1, is located at a distance from points A12 and A2 equal to the radius of the observable universe, rO (13.978 billion light-years, according to NASA's WMAP satellite). When we look at both points we see them as they looked 13.978 billion years ago. In both cases, the images we obtain in A1 correspond to the Cosmic Microwave Background (CMB), located at 13.978 billion light years, rO. Therefore, the three points are separated from the CMB by three converging lines measuring the same length, rO.

---

[4] https://es.wikipedia.org/wiki/Energ%C3%ADa_oscura

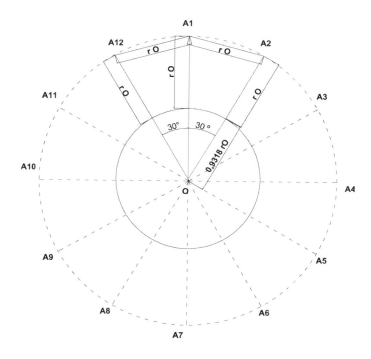

Figure 12 Information to obtain dimensions of the universe

In blue color is shown the information available from astronomical observations, necessary to determine the current dimensions of the universe. In red, the prolongation of the lines to the apex of the 30° angles of the two isosceles triangles, which allows to locate the center of the universe. rO = Radius of the observable universe

Consequently, at point A1 we will have two adjacent isosceles triangles at whose final coincident vertex will be the center of the universe. Both triangles evidently have an angle of 30° at the apex located at the center of the universe. In total coincidence with the theoretical value calculated in Chapter 3.

Like in Chapter 3, we obtain the radius of the sphere of the CMB, using the trigonometric functions of the angle of 15°:

$$Sin(15º) = \frac{1}{2}rO/(rO + x)$$

$$Sin(15º) = \frac{1}{2\sqrt{2+\sqrt{3}}}$$

Therefore:
$$x = rO\left(\sqrt{2+\sqrt{3}} - 1\right)$$

$$x = 0{,}93185\ rO \quad (E3\text{-}1)$$

This confirms the theoretical result of chapter 3. Figure 13 shows the complete plane of the model of the current universe in two dimensions where 12 points separated at the distance of the radius of the observable universe are observed rO divergent at an angle of Θ = 30°. As this scheme is equal in all three dimensions, it follows that they are a total of n=56 points 30° apart on the sphere of the universe.

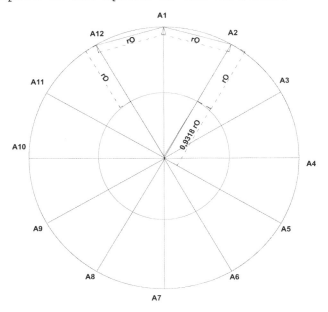

Figure 13 Scalable model of the universe 2D

rO = Radius of the observable universe (Length in light years traveled by light from the creation of the Universe to the date of observation.

The universe in three dimensions, is a sphere of radius equal to 1.93185 rO, with a spherical center of vacuum of radius 1.93185 **rO**. From the model we obtain the formula to size the universe at any time. Therefore:

$$rO = radius\ of\ observable\ universe$$

Diameter of the universe

$$Du = (1.93185) x\ 2\ rO$$

$$Du = 3.8637\ rO \qquad (E3\text{-}2)$$

Volume of the universe:

$$Vu = \frac{4}{3} x\ \pi\ x(1{,}93185\ rO)^3$$

$$Vu = 30{,}2x\ rO^3\ (light\ years)^3 \ldots\ (E3\text{-}3)$$

Dimensions of the Universe in the present epoch:

Assuming the age of the current universe is 13.978 billion years:

$$rO = 13{,}978\ x\ 10^9\ \text{ligth years}$$

$$Du = 3{,}8637 x 13{,}978 x 10^9 light\ years$$

$$Du = 54{,}006\ x10^9\ light\ tears$$

$$Vu = 30{,}2\ x(13{,}978\ x10^9)^3 (light\ years)^3$$

$$Vu = 82{,}478\ x10^{30}(light\ years)^3$$

5.1 New concept of the Observable Universe

Since space-time and matter are in the outer layer of the universe, the observable universe will be a portion of the area of the outer sphere of the universe. This means that the observable universe is equal to the area of a spherical cap, of the total sphere of the universe. Let's look at Figure 14, where we have demarcated in blue the spherical cap of the observable universe.

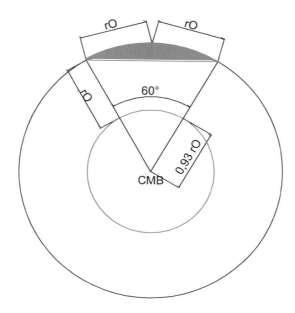

Figure 14 Sphere of the universe with the spherical cap of the observable universe

The area of a spherical cap is determined by the radius of the sphere "R" and the depth "h" of the cap according to the equation:

$$Ace = 2\pi R x h$$

In this case $R = 1,93185 \, rO$; y $h = rO \, x \, sen(15º) = 0,2588 \, rO$

So the area of the observable universe will be:

$$Auo = 2 \pi \, x \, 1.93185 \, rO \, x \, 0.2588 \, rO$$

$$Auo = \pi \, rO^2$$

The total area of the universe will be equal to the total area of the sphere, i.e.:

$$Atu = 4\pi (1,93185 \, rO)^2$$

$$Atu = 14.928 \, \pi \, rO^2$$

The percentage of the universe we can observe will be:

$$\frac{Auo}{Atu} = \frac{\pi \, rO^2}{14,928 \pi \, rO^2} = 0,066987 = \mathbf{6,6987\%}$$

Surprisingly we can only observe less than 7% of the universe

But what we can see from the universe doesn't have a quantitative-only response. The problem is more complex and interesting as other factors such as time, distance and vision cone are involved.

Vision acts in the form of a cone. The maximum field of view is located on the horizon as far away as possible. The same goes for the observation of the universe. The further we focus the greater the field of vision. This means that the older the image we observe of the universe, the greater the area observed. However due to the spherical shape of the universe the vision cone has curved edges. It's a kind of bulb, similar to a hood or the top of a projectile. The matter is observed housed only on the periphery of the bulb, stratified over time, in concentric rings along the bulb

In Figure 15, it is indicated, the spherical cap of the observable universe for three epochs: 1) Current Universe 13,978 x 10^9 light-years; 2) Universe for 6,989 x 10^9 light-years and 3) Universe 380.000 light-years, when light appeared (Image we observed CMB). Also shown in Figure 15 is the vision bulb.

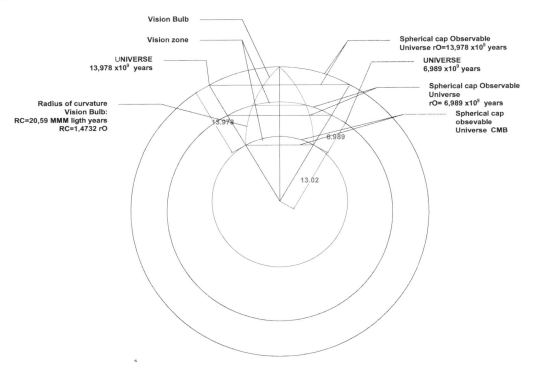

Figure 15 Spherical cap of the universe observable at different epochs

The shape of the vision bulb and the location of the images of the matter of the universe in concentric rings stratified in time along the vision bulb can be checked with an observational experiment, which we set out below:

In Figure 16, galaxies B and C are far from the Milky Way at a distance each of 4.61×10^9 light-years and galaxies A and D are located at 9.19×10^9 light-years. As the light coming from galaxies B and C takes 4.61×10^9 years to reach us, the image we can see corresponds to its location when it lacked 4.61×10^9 light-years to complete its trajectory. Similarly, as light from galaxies A and D takes 9.19×10^9 years to reach us, the image we can see corresponds to its location when it lacked 9.19×10^9 light-years to complete its trajectory. In Figure 16, we observe that the shape of the Vision Bulb and the location of images of galaxies in concentric rings stratified by eras of the universe around the vision Bulb are fully verified. This characteristic of the observable universe is of great use for the interpretation of the astronomical observations

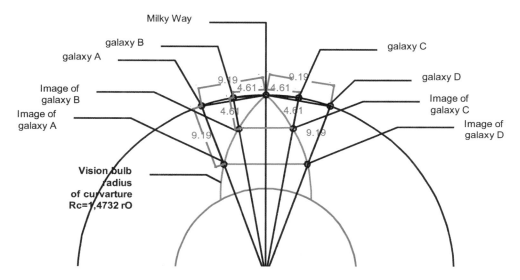

Figure 16 Figure Location of images of four galaxies in the viewing bulge

6   THE CREATION OF THE STARS AND GALAXIES OF THE UNIVERSE.

In the previous chapter we demonstrated, according to the equations of Einstein's Theory of Special Relativity, that the primal matter of infinitesimal dimensions, as a result of the energy created by its displacement at the speed of light, is duplicated creating a succession of clones, also at the speed of light. These clones of primeval matter were located at the almost infinite points of the space-time spherical caps created on each Miniverses. As we explained earlier, the mass cloning phase of primeval matter ends up becoming tangent to the fictional spheres of Miniverses.

From that moment on, clones of primeval particles whose mass was entirely hydrogen, begin their growth due to the energy derived from their displacement at the speed of light according to the equation: E1-3): $x = (dx + c\,t)$; $y = (dy + c\,t)$; $z = (dz + c\,t)$. With their growth they were transformed into primitive stars, which due to the absence of metalicity and even helium, were the largest and least life cycle developed in the history of universe. Due to their huge size the first stars at the end of their cycle, they made a great implosion of their nucleus, resulting in each of them a hypernova and a supermassive black hole. In this way the almost infinite points of space time of the spherical caps of all 56 Miniverses containing clones of the primal matter were transformed into galaxies made up of a supermassive black hole and a Molecular Cloud composed mostly of hydrogen and helium. Shortly before the formation of galaxies when the 56 space-time platforms of the Miniverses became tangent and the universe was created as a whole, the 56 primal particles located in the center of each Miniverse collapsed their mass directly, forming a primordial black hole.

Therefore, the previously unknown creation of the almost infinite galaxies of the universe is clarified. This theory of the origin of galaxies is corroborated and in turn explains the following deductions of what is observed from the CMB:

1) With the creation of the universe the light became visible
2) The primal universe was made up of a matter soup composed of hydrogen and helium.

We know that what we observe in the CMB really corresponds to the primal universe, and as we have explained before the formation of the multiple galaxies occurred at the beginning of the universe. So at that moment began, in the 56 Miniverses the process of exploding gigantic hypernovae, creating an incredible and almost infinite glare. It was a true cataclysmic event, a chain reaction, of hypernova explosions at the speed of light, in the same sequence that clones of primeval matter were created. This extraordinary event explains the sudden appearance of light in the primal universe and the beginning of the process of formation of all galaxies in the universe. This would consist, as we mentioned earlier, of a supermassive black hole and a molecular cloud of hydrogen and helium, with low metallicity

As a product of the nearly infinite numbers of hipernova, the entire universe was flooded with an immense Molecular Cloud of hydrogen and helium. Creating a "cosmic soup basically of hydrogen and helium". The almost infinite number of supermassive black holes created in the process and the 56 primordial black holes created a little earlier, were hidden as they could not be seen by the human eye and therefore an image that could be obtained from the primal universe, only hydrogen and helium soup would be observed.

The theory of the standard Cosmological model is coincidental to what would be observed of the primal universe. But it is very different in what according to that theory preceded that "hydrogen and helium soup".

According to the standard Cosmological model theory, in the early universe there are no black holes or primordial holes, not even stars, and hydrogen and helium do not come from Molecular Clouds produced by hypernovae. According to the standard Cosmological Model theory due to the energy of the Big Bang explosion, the mass of singularity disintegrated completely even at the molecular level, disappearing the atoms and their components. For this reason there was no remaining energy from that of the explosion. Subsequently the Standard Cosmological Model states that matter is reborn from the most elementary subatomic particles, creating the first atoms, then the first hydrogen molecules and after helium.

There is no information in the Standard Cosmological Model theory that scientifically explains this creative process, nor of the energy needed to create elements such as hydrogen and helium, in the immense amounts that formed the hydrogen and helium clouds of the early universe

6.1 Late Anisotropy confirms the theory of this research on the creation of primitive stars and galaxies.

However, the CMB observations clearly confirm the process of formation of galaxies through the hypernova chain and the creation of black holes described by the Theory of the New Cosmological Model, in what is called **late anisotropy**. Let's see what Wikipedia says about that period of the formation of the universe[5].

Late time anisotropy

Since the CMB came into existence, it has apparently been modified by several subsequent physical processes, which are collectively referred to as late-time anisotropy, or secondary anisotropy. When the CMB photons became free to travel unimpeded, ordinary matter in the

---

[5] https://en.wikipedia.org/wiki/Cosmic_microwave_background

universe was mostly in the form of neutral hydrogen and helium atoms. However, observations of galaxies today seem to indicate that most of the volume of the intergalactic medium (IGM) consists of ionized material (since there are few absorption lines due to hydrogen atoms). This implies a period of reionization during which some of the material of the universe was broken into hydrogen ions.

The CMB photons are scattered by free charges such as electrons that are not bound in atoms. In an ionized universe, such charged particles have been liberated from neutral atoms by ionizing (ultraviolet) radiation. Today these free charges are at sufficiently low density in most of the volume of the universe that they do not measurably affect the CMB. However, if the IGM was ionized at very early times when the universe was still denser, then there are two main effects on the CMB:

1. Small scale anisotropies are erased. (Just as when looking at an object through fog, details of the object appear fuzzy.)
2. The physics of how photons are scattered by free electrons (Thomson scattering) induces polarization anisotropies on large angular scales. This broad angle polarization is correlated with the broad angle temperature perturbation.

Both of these effects have been observed by the WMAP spacecraft, providing evidence that the universe was ionized at very early times, at a redshift more than 17.[clarification needed] The detailed provenance of this early ionizing radiation is still a matter of scientific debate. It may have included starlight from the very first population of stars (population III stars), supernovae when these first stars reached the end of their lives, or the ionizing radiation produced by the accretion disks of massive black holes.

The time following the emission of the cosmic microwave background—and before the observation of the first stars—is semi-humorously referred to by cosmologists as the Dark Age, and is a period which is under intense study by astronomers (see 21 centimeter radiation). Two other effects which occurred between reionization and our observations of the cosmic microwave background, and which appear to cause anisotropies, are the Sunyaev-Zel'dovich effect, where a cloud of high-energy electrons scatters the radiation, transferring some of its energy to the CMB photons, and the Sachs-Wolfe effect, which *causes photons from the Cosmic Microwave Background to be gravitationally redshifted or blueshifted due to changing gravitational fields*

The above quote clearly states that the ordinary matter of the primal universe was made up of hydrogen and helium and its origin was linked to the appearance in the primal universe of supernovae and massive black holes in total accordance with our theory

Recent astronomical observations[6] made by researchers who have used the ALMA telescope located in Atacama, Chile; confirm the co-evolution of galaxies and supermassive black holes in the early universe, 13,100 million years ago. Evidence that contradicts the theory of the formation of supermassive black holes by multiple junctions of IMBH intermediate-mass black holes supported by the Standard Cosmological Model and fully coincides with the theory on the formation of galaxies and supermassive black holes supported by the New Cosmological Model, presented in this chapter.

Sharp images from the James Web Supertelescope (JWST) reveal the existence of galaxies in the universe even earlier, just 320 million years after the Big Bang[7]. These observations have led scientists to question the Standard Cosmological Model and to raise the need to study new cosmological models.

7 TIME IS CREATED BY THE MOVEMENT OF MATTER

Once the initial stage is concluded with the creation of the first model of the universe, its expansion as a whole will begin which, as we have said before, is in the three dimensions at the speed of light.

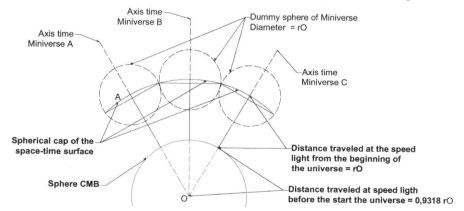

Figure 17 Trajectory and surface axes of space-time in the Miniverses

---

[6] Detectan colosal tormenta de agujero negro en el universo primitivo, la más antigua hasta la fecha | Ciencia y Ecología | DW | 18.06.2021

[7] Hito histórico: localizaron la galaxia más lejana en el Universo (clarin.com)

Because the radius of the sphere of the universe is very large in relation to the radius of the space-time cap of each Miniverse, for practical purposes, we will consider that flat surface with two physical dimensions and a third dimension which is time, which is parallel to the translation axis of each of the Miniverses.

As the perception of any fact is relative according to the location of the observer, we will see the process of expansion: first as an observer located at the center of the universe, the site where the Big Bang exploded. Point "O" illustration 17 and secondly, as an observer located on the space-time platform of any of the 56 Miniverses. Point "B" Figure 17

1. For an observer located at the center of the universe will perceive, that the 56 platforms of space-time (PET) are moving away simultaneously in the 56 directions at the speed of light and will also perceive that the PET of all the miniverses expand simultaneously, at the speed of light.

2. An observer located in a PET of one of the miniverses (Example on Earth), will not perceive anything at all. Since it has no reference point to perceive the displacement of the PET that contains it, nor will it be able to perceive the expansion of the PET at the speed of light, because it does not have a reference point to perceive it, since it and the environment that surrounds it also grows or expands at the speed of light.

We've looked at what happens to movement because of expansion. We will also look at what happens over time. We will take again the two points of observation above and in addition to the movement, we will observe how time is perceived.

1. For the observer located at the center of the Universe, he perceives that there are 56 independent time axes since each PET has a divergent trajectory and as the axes of time run parallel to its trajectory, there will be a different time in each PET. But something extraordinary happens, even if the axes of time and motion are not coincident, they are synchronized. That is to say that if there were a clock in each PET marking the time since the birth of the universe, it would mark in all 56 cases the same time. This is a relevant fact of extraordinary importance. We can say that movement and time are linked phenomena and that they always exist together and that there is an official time throughout the universe and it is the same for each PET being synchronized all in perfect harmony since the explosion of the Big Bang. And

something even more decisive, the official time of the universe passes at the speed of light and therefore is unalterable and inviolable.
2. For the observer located on the space-time platform of one of the miniverses, time does not seem to elapse as he cannot perceive the displacement movement of the platform, nor of its growth. Your perception of time will be only by the time created by the movement that may exist between the objects located on that platform of space-time. There are three causes that create movement and therefore perception of time for observers located on space-time platforms:
   a. The movement of bodies located in their environment, due to the gravitational pull force.
   b. The transformation of matter into energy or energy into matter, which may occur in its environment (explosion of supernovae, creation of red giants, etc.)
   c. The extra natural movement, produced by man, with its space vehicles.

This is a clear demonstration of the relativity of the perception of space and time, something that was known since Galileo and that Einstein could explain differently, in his famous theory of Relativity. However, in both theories both Galileo and Einstein had as a scenario a static universe, which was not exposed to the complex process of expansion that we have described before. Therefore both theories should be revised considering the repercussions generated by the expansion of the universe

7.1 Time does not dilate with speed

One of the most surprising and "proven" conclusions of Albert Einstein's theory of Special Relativity is the supposed expansion of time with speed.

For the calculation of the trajectory of a mobile, in space-time, the Standard Cosmological Model uses the equation developed by Minkowsky ($ds^2 = dx^2 + dy^2 + dz^2 - c^2t^2$). That equation subtracts time, which means that space shrinks over time in clear disagreement with the expansion of the universe. The Minkowsky equation aims to show that the invariance of the speed of light is fulfilled at the expense of the deformation of time. It is convenient to remember that at the time that the 2 theories of Einstein's Relativity were formulated (Special 1905 and General 1915), the theory of the expansion of the universe was unknown.

In the Theory of the New Cosmological Model the relationship between space and time is clearly defined and the corresponding equations have to consider, due to the expansion of the universe, that the variables of space (x, y, z) increase over time at the speed of light, and that in addition the platform of space-time where matter resides is permanently transferred at the speed of light , on the z-axis of hyperspace parallel to the time axis. Due to the complexity of this process two equations are required.

(E 7-1) $\quad x = (dx + c\,t)\;; y = (dy + c\,t)\;; z = (dz + c\,t); Z = ct$

(E7-2) $\quad ds^2 = \left(dx + \frac{v}{c}dx\right)^2 + \left(dy + \frac{v}{c}dy\right)^2 + \left(dz + \frac{v}{c}dz\right)^2 + c^2 t^2$

The first equation measures how the space-time platform increases and moves with universal time and is similar to the equation (E1-3) applied to matter. The second equation measures the length of a trajectory at a certain speed.

And it is fulfilled:

$(dx + c\,t) = \left(dx + \frac{v}{c}dx\right)\;; (dy + c\,t) = \left(dy + \frac{v}{c}dy\right); (dz + ct) = \left(dz + \frac{v}{c}dz\right)$

Likewise in Minkowsky geometry both equations are quadrivectorial, in which there are three spatial coordinates and a temporal one that is time, but with notable differences, which we indicate below

Since each movement generates a time, there are two types of movement and time. The time of movement of the translation of objects in the planes formed by the x/tx, y/ty axes; z/tz, located on the surface of space-time and Universal Time corresponding to the movement of objects in Z hyperspace (parallel to the axis of time). The movement of the objects in the first case, planes x/tx, y/ty, z/tz is at the nominal speed, the part corresponding to the distances dx, dy and dz. And at the speed of light the corresponding to the increase of the distance in those axes due to the expansion of the universe **ct**. In the second case the movement in Hyperspace Z, parallel to the axis of time will also be at the speed of light, equal to **(ct)**. The result of a given movement will be the sum of both spaces traveled and the times used, in both areas.

It must be understood that the platform of space time moves at the speed of light and also the space and objects they contain grow at the speed of light. Therefore when an object moves from an "a" point to a point "b" on the x-axis at a v-speed, the initial space between both points increases in time at the speed of light and in addition that object moves orthogonally by the movement of the platform of space-time at the speed of light.

All this complex movement can be represented for various speeds on the Figure 18 chart, and the results in the calculation table 2

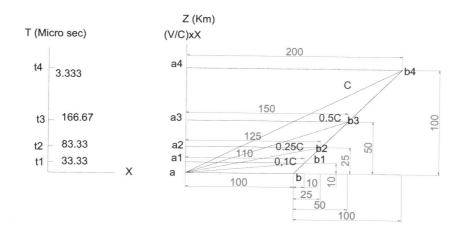

Figure 18 Velocities and trajectories in space-time

This graph shows the movement of an object moving from point "a" to point "b", at 4 different speeds, indicated as fractions of the speed of light. Five space-space platforms are represented, the measurements of which are made under the space-time platform metric corresponding to the start of the movement:

- Space-time platform (STP, a-b). It corresponds to the dimensions and location of the platform at the beginning of the movement.
- Space-time platform (STP, a1-b1). Corresponds to the dimensions and location of the platform at the end of the movement at the speed 0.1C
- Space-time platform (STP, a2-b2). Corresponds to the dimensions and location of the platform at the end of the movement at the speed 0.25C
- Space-time platform (STP, a3-b3). Corresponds to the dimensions and location of the platform at the end of the movement at the speed 0.5C
- Space-time platform (STP, a4-b4). Corresponds to the dimensions and location of the platform at the end of the movement at the speed 1C

Total distance of the path ds, is the distance a-bn, in blue. It is calculated using the equation (E7-2)

$$ds\,(0.1C) = \sqrt{(dx + 0.1dx)^2 + (0.1dx)^2}$$
$$ds\,(0.25C) = \sqrt{(dx + 0,25dx)^2 + (0,25dx)^2}$$
$$ds\,(0.5C) = \sqrt{(dx + 0,5dx)^2 + (0,5dx)^2}$$
$$ds(1C) = \sqrt{(dx + dx)^2 + (dx)^2}$$

In relation to time as we have two axes of motion and time (X axis and Z axis), the sum will also be vector:

$$Tt = \sqrt{\bigl(tx + t(xc)\bigr)^2 + (TuZ)^2} \quad (E7-3)$$

Where:

$$Tt = \text{total movement time}$$

$$tx = \text{time traveled in x-axis at rated speed}$$

$$tx = dx/v$$

$$t(xc) = \text{time traveled on x-axis at the speed of light}$$

$$t(xc) = \left(\tfrac{v}{c}dx\right)/c = t$$

$$Tuz = \text{Universal time traveled in hyperspace } Z$$

$$Tuz = t$$

With regard to speeds we have:

$$Vnmc = \text{Speed according to the New Cosmological Model}$$

$$Vncm = ds/Tt \quad (E7-4)$$

$$Vetr = \text{Speed according to Einstein's theory of Relativity}$$

$$Vetr = \frac{dx}{tx + t(xc)} \quad (E7-5)$$

In the case of the Theory of the New Cosmological Model, the totality of movements and times is perceived both on the **x**-axis and in hyperspace **Z**. In the case of Einstein's Theory of Relativity, observers located on earth perceive only movement and time on the x-axis. They cannot perceive by their location the movement and time in hyperspace Z.

These results in observers located on Earth, coming to the wrong conclusion that time is extended since at the time of travel on the x-axis at the nominal speed, the time of the fraction of the x-axis that travels at the speed of light is added due to the expansion of that axis. As the metric of the universe changes at the speed of light, the observer perceives that the distance is the same. 100 km, as it does not perceive the increase of space due to the way he and his environment also grow at the speed of light. See Table 2:

Table 2
SPEED TABLE

1	2	3	4	5	6	7	8	9	10	11	12
v	dx	dZ	dcx	dx_2	ds	tx	t(xc)	TuZ	Tt	Vncm	Vtre
Km/sec	Km	Km	Km	Km	Km	Mcr sec	Mcr sec	Mcr sec	Mcr sec	Km/sec	Km/sec
30.000	100	10	10	110	110,45	3.333,33	33,33	33,33	3.366,82	32.805,39	29.702,97
75.000	100	25	25	125	127,48	1.333,33	83,33	83,33	1.419,11	89.831,03	70.588,24
150.000	100	50	50	150	158,11	666,67	166,67	166,67	849,84	186.045,96	120.000,00
300.000	100	100	100	200	223,61	333,33	333,33	333,33	745,35	300.000,00	150.000,00

Column 1: v, Nominal speed. **Column 2**: dx, Distance between points at the beginning of the movement. **Column 3**: dZ, Distance traveled on the Z-axis of hyperspace, bounded in magenta in Figure 18. **Column 4**: dcx, Distance traveled on the x-axis at the speed of light, bounded in red in Figure 18. **Column 5**: dx_2, New distance between points "a" and "b" at the end of the movement, bounded in green in the diagram in Figure. 18. **Column 6**: ds, Actual distance traveled in space-time, vector sum of columns (2+4) and 3, represented as a blue line in the Figure 18 graph. **Column 7**: tx, Travel time on the x-axis at the nominal speed; values of column 2 divided by the values of column 1. **Column 8**: t(xc), Time traveled on the x-axis at the speed of light; values of column 4 divided by the speed of light. **Column 9**: TuZ, Time in hyperspace, time axis of the universe; values of column 3 divided by the speed of light. **Column 10**: Tt, Total time, vector addition of columns 7, 8 and 9. **Column 11**: Vncm, Speed according to the NCM; values of column 6/column 10. **Column 12**: Vetr, Velocity according to the ETR, apparent velocity on the x-axis; column 2/ (column 7 + column 8).
The results reflected in the table 2 reveal the following:
- The observers located on Earth, appreciate the movement on the x-axis at the nominal velocity dx, being for them the distance between "a" and "b", the same before and after the movement, that is, the values in column 2. They also perceive the movement time at the

nominal speed tx, values of column 7, and the time traveled on the x-axis at the speed of Light t(xc), values of Column 8.

- The value of the velocity that these observers perceive, accordingly is: Vetr, the quotient of column 2 divided by the sum of columns 7 and 8 dx/(tx+t(xc)), which is reflected in column 12.
- The values of the resulting speeds in column 12 are lower than those of the nominal velocities indicated in column 1. These observers, as they do not have the vision of hyperspace, do not know that the trajectory increased during the journey, and therefore arrive at the erroneous conclusion that the decrease of the speed is because the time has been dilated during the journey.
- Observers located in the center of the universe clearly note that there is no dilation of time. The real value of velocity, Vncm, is reflected in column 11, whose values correspond to the quotient between the values of column 6 and the column 10, which represent the total distance of the ds route and the total time Tt respectively. These results are totally consistent with the laws of physics.
- Special relevance has the example of moving a mobile at the speed of light, since according to the ETR the speed by the effect of the apparent dilation of time is reduced to 50% (150,000 KM/sec), while in the NCM, the value is 300,000 km/sec in clear concordance with the constant value of the speed of light, which is a confirmatory evidence of this new theory.

The results of this experiment have a huge and transcendental importance: The following theory is fully demonstrated without any doubt.

7.1.1 "In an expanding universe, time is not distorted by movement at high speeds"

The concept of time distortion formulated in Einstein's Theory of Relativity is wrong. Time does not dilate with speed, what happens is that the trajectory is lengthened, due to the expansion of the universe.

8 SCIENTIFIC EXPLANATION OF HUBBLE'S LAW AND PROOF OF THE EXPANSION MODEL OF THIS RESEARCH

Figure 19 shows the universe, according to our model, depending on the radius of the observable universe **r0**. We have placed the galaxy **b** on the surface of space-time on the time axis **B**, galaxy **c** on the surface of space-time on the time axis **C** and galaxy **d** on the surface of space-time on the time axis **D**.

We can calculate the rate of separation of those galaxies from galaxy **"e"** located on axis **E,** our point of observation and therefore it is the Milky Way.

$$V = espace/time$$

$$T = \frac{rO}{C} \quad (E8-1)$$

$$Vb = \frac{0{,}75\, rO}{\frac{rO}{C}} = 0{,}75\, C$$

$$Vc = \frac{0{,}5\, rO}{\frac{rO}{C}} = 0{,}5\, C$$

$$Vd = \frac{0{,}25\, rO}{\frac{rO}{C}} = 0{,}25\, C$$

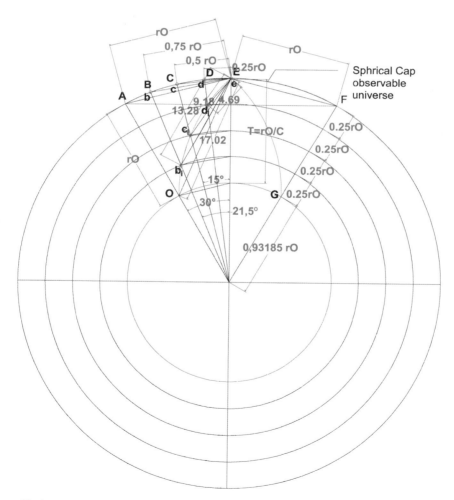

Figure 19 Universe depending on the radius of the observable universe

Therefore it must be complied with under Hubble's law[8]:

$$\frac{speed\ b}{distance\ (b-e)} = \frac{speed\ c}{distance\ (c-e)} = \frac{speed\ d}{distance\ (d-e)}$$

Replacing values:

$$\frac{0{,}75\ C}{0{,}75 rO} = \frac{0{,}50 C}{0{,}50 rO} = \frac{0{,}25\ C}{0{,}25 rO} = \frac{C}{rO} \quad (E8-2)$$

The result clearly reveals the full and undoubted compliance with the constant proportionality between speed and distance of separation of galaxies established in Hubble's law.

Therefore in our model of expansion of the universe, in which the platform of space time grows and moves at the speed of light "C" is fully met, the ratio of constant proportionality between the speed and distance between the galaxies observed by Hubble. It is worth clarifying that what we are measuring is the speed at which the space between the two galaxies is expanded and not a movement typical of them. Specifically, it's about the rate of separation of the time axes where galaxies are housed. Due to the change of the metric to the speed of light the relative distance between the time axes is constant.

In our model the Miniverses are kept permanently apart at the distance from the radius of the observable universe **rO**. Because the radius of the observable universe grows permanently at the speed of light, the speed of separation of the miniverses' axes will always be equal to "C" the speed of light. In the example the axes corresponding to the Miniverse are A; E and F. The angle of separation between the adjoining Miniverses is 30°, therefore the axes A and F. move away from the E axis at the speed of light, while the intermediate axes do so at a fraction of the speed proportional to its distance to the D axis. We can also conclude that the rate of separation of two bodies, as a result of the expansion of the universe, will be equal to a fraction of the speed of light equivalent to the fraction of the separation distance between the bodies, relative to the radius of the observable universe. Therefore you must always comply:

$$X = distance\ (a-b)$$

$$Vab = \frac{X}{rO} \times C \quad (E8-3)$$

This proportionality between distance and speed of separation, from galaxies, could be seen by Edwin Hubble in his famous observations made at the

---

[8] https://en.wikipedia.org/wiki/Hubble's_law

beginning of the last century at the Mount Wilson Observatory, located in the vicinity of the city of Pasadena, California. These observations, subsequently, were the basis for the formulation of the theory of the expansion of the universe and the Bing-Bang theory.

Previously we found that the proportionality between the speed and distance of galaxies: **b**, **c**, and **d**, in Figure 19, has a constant value equal to **C/r0**. However that was not what Hubble observed, as I could not see the galaxies located in the spherical cap of space time of the observable universe. Hubble was only able to observe the past images of the aforementioned galaxies. Therefore we should check whether the past images of the b_i, c_i and d_i galaxies also meet the constant speed/space ratio:

The images of the galaxies in the example in Figure 19 are located at different distances from the point of observation and therefore at different times of the universe. Previously we will take the distances to a fraction from the radius of the observable r0 universe of the current era. In Figure 19, it has been used as a value of **r0**=13.98 on a drawing scale of 1=10^9 light-years. The values of the speeds, we obtain them by dividing the distance as a fraction of the radius of the observable universe between time **T**= **r0/C**. The results are shown in the Table 3:

Table 3
Speed/distance

Observed point	Age of the universe of the observed point	Distance between the point of observation and the observed point	Distance/r0 r0=13.98	Speed	Speed/ Distance
O	0	17,02	1,217 **r0**	1,217 **C**	**C/r0**
b_i	0,25 **r0**	13,28	0,950 **r0**	0,950 **C**	**C/r0**
c_i	0,50 **r0**	9,18	0,656 **r0**	0,656 **C**	**C/r0**
d_i	0,75 **r0**	4,69	0,335 **r0**	0,335 **C**	**C/r0**

From the results table we can see that in all cases regardless of the age of the universe the speed/distance ratio remains constant = **C/r0**. Therefore, the Hubble law speed/distance ratio is verified to be an independent constant from the time it is measured. This leads us to two important conclusions:

1) The difference obtained in the measurements of the Hubble H_0 constant, according to the data of the early universe, with respect to the data

obtained from the recent universe, is due to differences in the margin of error of the data obtained or differences in the methods used.

2) The very important conclusion that the expansion of the universe is not accelerated, and the confirmation of its expansion at the constant speed of light, as established by the Theory of the New Cosmological Model.

We have included in the table in addition to galaxies, point **O**, whose distance to observation point "**E**" is 17.02 on the scale of the plane and therefore by dividing the distance by the radius of the observable universe (13.98) a value of 1,217 **rO**. The value looks disconcerting, since the distance 1,217 **rO**, is greater than the radius of the observable universe and its speed of distance from point "**E**", v(O-E)=1,217 **C**, is higher than the speed of light. It seems like a double mistake because supposedly as far away as we can observe in the universe is an object that is at the distance from the radius of the observable universe **rO** and nothing can travel faster than the speed of light **C**. But it's not really a mistake as we'll explain below:

Point **O** belongs to the axis of time **A** and is located at the beginning of the universe, at a distance equal to the radius of the observable universe **rO** relative to point **A** on the surface of space-time on axis **A**. Therefore the OA distance constantly increases at the speed of light. Which means there's no error, just change the observation point? "The angular divergence of the time axes from which the observation is made creates the difference and produces the illusion of an accelerated movement since we are measuring the speed from the time axis where we are located and not on the axis parallel to the movement".

This angular difference that allows to visualize distances greater than the radius of the observable universe is completely consistent with the model of the shape and dimensions of the universe according to the Theory of the New Cosmological Model, which we will go on to demonstrate:

In Figure 19, the distance of the **O-E** line, which in turn corresponds to the distance of the image from the galaxy ai to observation point **E**, previously calculated with a length of 1,217 **rO**, is also the uneven side of an isosceles triangle whose other two equal sides measure **rO** the radius of the observable universe and are joined by an angle of 75°. So using trigonometry we have:

$$Distance\ (O - E) = 2\sin(37.5º)rO \quad \textbf{(E8 – 4)}$$

$$Distance\ (O - E) = 1{,}2175\ rO$$

Resulting in exactly the same previously calculated value. What demonstrates the accuracy and coherence of the model of the universe, of the Theory of the New Cosmological Model.

There is another important conclusion, galaxies are observed from the earth at a distance greater than they actually are, and they are also perceived to move away at a speed proportionally greater than their true distance rate. This difference between the actual and observed values is increased proportionally to the distance at which they are from the observation point. To illustrate the above statement, let's look at the comparison chart of the distances and speeds of galaxies and that of their observed images (Table 4):

Table 4
Real and observed distances and speeds

Galaxy	Actual distance	Actual Speed	Observed distance	Observed speed
b	0,75 r0	0,75 C	0,950 r0	0,950 C
c	0,50 r0	0,50 C	0,656 r0	0,656 C
d	0,25 r0	0,25 C	0,335 r0	0,335 C

To calculate the age of the universe based on the Hubble constant we must answer the following two questions in advance:
1) What value should be taken to calculate the age of the universe between H0 70Km/sec/Mpc corresponding to measurements of the ancient universe or 73.5Km/sec/Mpc of measurements of the current universe?.
2) What are we measuring and how do we relate it to the age of the universe?

First we'll answer the second question:
What are we measuring and how do we relate it to the age of the universe? Does the answer involve other questions: Depends on the shape of the universe and the speed of its expansion?

However, the Theory of the Standard Cosmological Model solves this problem, without previously defining the shape of the universe and its speed of expansion, and calculates the age of the universe simply by reversing the value of Hubble's constant.

$$Age\ of\ the\ universe = \frac{1}{H_o} = \frac{1 Mpc}{\frac{70 Km}{sec}} \quad (E8-5)$$

$$\text{Age of the universe} = 3.2616 \frac{10^6 ly}{\frac{70}{C}} = 13.978 \, x10^9 \text{years}$$

In the case of using the H_0 value corresponding to the early universe, or using the value of 73.5 Km/sec/Mpc, in which case the age of the universe would be:

$$\text{Age of the universe} = 3.2616 \frac{10^6 ly}{\frac{73.5}{C}} = 13.312 \, x10^9 \text{years}$$

This means that with whatever the chosen value of H_0 the implicit form of the Universe of the Standard Cosmological Model is a sphere with a radius equal to the radius of the observable universe and its speed of expansion is equal to the speed of light. This is a contradiction because it matches the dimensions of what the same theory calls the sphere of only the observable part of the universe. And on the other hand, by assuming that the universe expands at the speed of light, the speed of its expansion ceases to be unknown.

The above reasoning leads us to a reflection. Hubble's constant cannot be used to measure the age of the universe without knowing its speed of expansion as we would have a single equation and two unknowns. Therefore its outcome would be indeterminate. On the other hand the expansion of the universe cannot be accelerated because in that case the time of the universe would also be accelerated, since both concepts expansion speed and time are directly related by a constant the constant of Hubble H_0. Accelerated time contradicts the concept of time and its possibilities of measuring it.

Additionally in a universe whose radius is the observable universe has no space for the inner sphere of the CMB Cosmic Microwave Background, which is an undeniable reality, being constantly studied, and especially in this case the study of Hubble's constant in the early universe, for which NASA's WMAP satellite was used, which took some of the data from the study precisely of the CMB. Undoubtedly one of the most relevant deficiencies of the Standard Cosmological Model, in our opinion, is the lack of a model of its form that allows it to size the universe, hence the explanation of this contradiction.

The Theory of the New Cosmological Model if it has a model of the universe that also contemplates its dimensions according to the radius of the observable universe. Therefore to calculate the age of the universe according to our model, we have to make use of the measurements of the Hubble constant according to what that measurement represents in the model and from there deduce the corresponding age of universe, considering that the universe expands at the constant speed of light.

To obtain the age of the universe, according to the model of the Theory of the New Cosmological Model, we must equal the speed/distance ratio, obtained according to the data taken from that model with the Hubble constant:

$$\frac{C}{rO} = H_o \quad (E8-2)$$

We must first choose the value of the Hubble constant. We will take the value supplied by the W MAP satellite, considering the accuracy of its instrumentation:

$$\frac{C}{rO} = \frac{70 km/sec}{1 Mpc}$$

From the above equation we can calculate the radius of the observable universe and consequently the age of the universe:

$$rO = \frac{1\ Mpc\ x\ C}{70 Km/sec}$$

$$rO = \frac{3{,}2616\ al\ x\ 3x10^5 \left(\frac{Km}{sec}\right)}{70\ Km/sec}$$

$$rO = 13{,}978\ x10^9 ligtyears$$

Consecuently:

Ratio of the observable universe $(rO) = 13{,}978 \times 10^9 ly$

Age of the universe $= 13{,}978 \times 10^9 years$

Below in Figure 20 we draw the universe with its current dimensions:

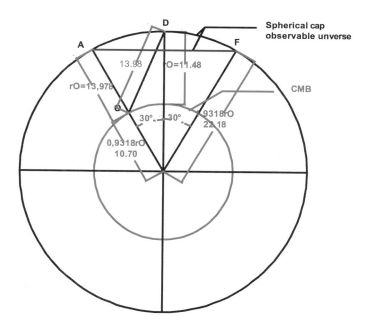

Figure 20 Image of the universe with its current dimensions. Scale 1:10⁹ light years

8.1 The expansion of the universe is not accelerated

We have seen that the universe expands at the constant speed of light. However, the Cosmological Standard Model establishes that the universe expands in an accelerated manner and uses as an argument that at "greater distance objects move away at a greater speed, being constant the Velocity/distance relationship", if providing evidence for that statement.

In order to clarify this controversy we will make a simulation where the opposite of what the CEM affirmed is demonstrated. Let's look at the figure 21

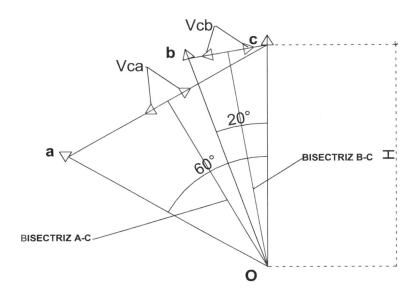

Figure 21 Mobiles at constant speed from the same source

In Figure 21 we see three mobiles: "a", "b" and "c" that start from the common point "o" with divergent angles with respect to the mobile "c". Mobiles move at the constant speed "v". We will evaluate the speed/distance ratio when the mobiles reach the distance "H" in their trajectory. The results are indicated in Table 5

Table 5

A	B	C	D	E	F	G	H	I	J
Mobile	speed	distance to origen	angle	angle bisector θ	sin θ	Distance from "c" (2 senθx *H*)	Vel. Away from "c" (2senθx *v*)	speed /distancia	speed/distancia
c	*v*	*H*							

b	***v***	***H***	20	10	0,1736	0,35***H***	0,35***v***	0,35***v***/0,35***H***	***v/H***
a	***v***	***H***	40	20	0,3420	0,68***H***	0,68***v***	0,68***v***/0,68***H***	***v/H***

In column "I" we can see that the speed/distance ratio is constant and the speed increases with the distance in the example where all bodies move at constant speed. Consequently the claim of the accelerated expansion of the universe held by the standard Cosmological Model based on the increase in the speed of distance with distance is a fallacy; which joins the multiple deficiencies of this cosmological model exposed in the preceding and subsequent chapters of this book.

We can check by this method the values previously obtained by our model

Table 6

A	B	C	D	E	F	G	H	I	J	K
Galaxy	Speed.	distance to origen	Angle	Bisector angle Θ	sin Θ	Dist. from "e" (2 senΘx Ru)	Vel. Away from "e" (2senΘx *c*)	Vel. Away from "e" froml origen (Col H*Ru)	Speed /distance	Speed /distance
e	*c*	1,93rO								
d	*c*	1,93rO	7,5	3,75	0,07	0,25rO	0,13 c	0,25 c	0,25c/0,25rO	c/rO
c	*c*	1,93rO	15	7,5	0,13	0,50rO	0,36 c	0,50 c	0,50C/0,50rO	c/rO
b	*c*	1,93rO	22,5	11,25	0,20	0,75rO	0,39 c	0,75 c	0,75c/0,75rO	c/rO

The table confirms the results obtained previously in this chapter. It is interesting to highlight the curious case of relativity between space and time that happens in this example. The results of the column "H" are multiplied by the radius of the universe 1.93rO to include the stretch from the origin of the Big-Bang, until when time begins.

8.2 Clarification on the value of Hubble's parameter and its variation over time

Hubble's constant is not really a constant in the orthodox sense of the word since we have defined it according to the equation:

$$H_o = \frac{C}{rO} \quad (E8-2)$$

As the value of the speed of the **C** light is constant and the **rO** value, the radius of the observable universe, permanently increases at the speed of light, the value of the Hubble constant permanently decreases at the speed of light. This means that the universe expands at the speed of light as indicated by our Theory of the New Cosmological Model. This seems like a contradiction but it is not, if we take a closer look at the problem and why and when we measure that parameter.

Hubble's parameter is measured to determine the age of the universe, according to its speed of expansion. Therefore, the rate of variation of the Hubble parameter value indicates the speed of expansion of the universe and its rate of aging. So it's obvious that it can't be a constant. This means that in a smaller, younger universe the value of Hubble's parameter would be proportionally higher. For example for the universe of the current half-time, approximately 7 billion years, it will be approximately double, i.e.: 140 Km/seg/1Mpc. But that value, which is only important for astronomical observations, is totally useless, because when the universe was 7 billion years old no one could make astronomical observations, since neither the earth nor the Sun existed.

The value of the Hubble parameter that is of interest is that corresponding to the time of existence and the age of the universe to the date of measurement, and since the current time of existence of the universe is very large value approximately 14 billion years, the radius of the observable universe **rO** 14 billion light years, has changed virtually nothing in the 100 years since the Hubble parameter is being measured. Therefore the Hubble H_0 parameter is currently, in practice, a constant and will probably remain constant for the rest of humanity's existence.

$$H_o = \frac{C}{rO} \cong Constant$$

8.3 Conclusions of the chapter of Hubble's law

- We have achieved the scientific explanation of Hubble's law after nearly 90 years of its formulation..
- The scientific explanation of Hubble's Law represents a strong confirmation of the Cosmological Model of Special Relativity.
- The universe is not expanding rapidly. It expands at the constant speed of light; which ratifies the discarding of the theory of Dark Energy.

9 THE CURVATURE OF SPACE-TIME, IN THE THEORY OF THE NEW COSMOLOGICAL MODEL

In Figure 22, we observe an infinitesimal particle cloned from primeval matter as it is implanted on the surface of space-time.

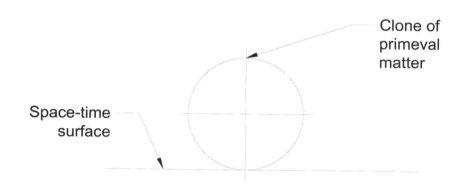

Figure 22 Clone of the primordial particle implanted at a point in space-time

In the process of expansion grow at the speed of light both space-time and the spherical mass implanted on its surface. As the spherical mass has a dimension parallel to Hyperspace Z and the plane of space-time does not, the following happens: When the spherical body grows on the Z axis it produces a curved deformation of the plane of space-time whose depth will be necessary so that the length of the curvature produced in that plane is equal to the radius of the spherical body. From that moment on, the curvature stops increasing and the spherical body and space-time, already curved, will continue to grow harmoniously and in a synchronized way. The angle of the circumference arc whose length is equal to the radius of the circumference corresponds to 1 radian

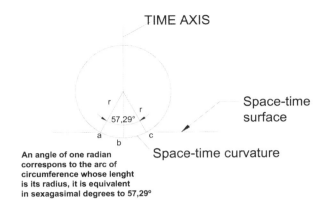

Figure 23 Spherical body with the curvature of space-time

In Figure 23 we see the curved deformation produced in the plane of space-time. As it is a spherical body the deformation is three-dimensional and corresponds to the spherical cap in Figure 24

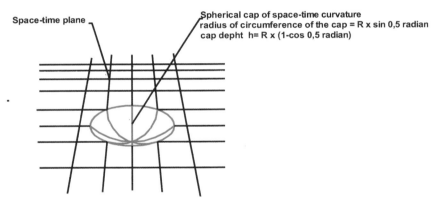

Figure 24 Spherical cap of the curvature of space-time

By simple trigonometric calculations we obtain the dimensions of the cap according to the radius of the spherical body (R_{sb}):

Spherical cap circumference radius (R_{scc}):

$R_{scc}= R_{sb} \times sin\ 0.5\ radian$

$R_{scc} = 0,4794\ R_{sb}$ (E9-1)

Cap depht (h) is:

$h=R_{sb} \times (1-cos\ 0.5\ radian)= 0,1224\ R_{sb}$

$h= 0,1224\ R_{sb}$ (E9-2)

All the celestial bodies of the universe: stars, planets, satellites, etc., will form a similar spherical cap, a kind of niche, on the surface of space-time which will expand at the speed of light in a harmonic manner and synchronized with the celestial body, so we then formulate the following theory:

9.1 The spherical cap of the curvature of space time

"Due to the expansion of the universe all celestial bodies: stars, planets, satellites, etc., create a curved deformation of space time in the form of a spherical cap, allowing a harmonic growth at the speed of light from the celestial body and the space-time platform where they are housed the same. The dimensions of the above-mentioned spherical cap in relation to the radius of the celestial body (Rcb) shall be: radius of the circumference of the spherical cap Rscc=Rcb x sine 0.5 radian, and the depth of the spherical cap h = Rcb x (1-cosine 0.5 radian)"

This theory is noticeably different from that of Einstein's General Theory of Relativity

Below are the following aspects of the curvature of space-time according to Einstein's Theory of General Relativity:

1) The curvature of space-time is caused by the mass of the bodies.

2) Space-time is curved where the massive body sits and also curves an area of the surrounding space-time, being the area of that area proportional to the mass of the massive body.

3) The complexity of its calculation is so extreme that only very simple cases around isolated stars have been solved.

4) Einstein dispenses with the concept of gravitational force in Newton's theory and replaces it with the effect of space-time deformation surrounding the massive body.

5) In the formation of the curvature of space time the effect of the expansion of the universe is ignored

In contrast, let's look at the above points on the theory of the curvature of space-time according to the NCM.

1) The curvature of space-time is related to the volume of the bodies and not their mass

2) It curves only the place where the celestial body sits. There is no distortion of space-time in the vicinity of the body

3) The calculation is very simple and can be easily obtained for any celestial body.

4) As there is no distortion of space-time in the vicinity of the celestial body, there is no geometric effect affecting another body.

5) The expansion of the universe is the cause of the formation of the curvature of space-time.

We have contrasted the differences between the two theories regarding the curvature of space-time. In this chapter we will only highlight those differences. In the next chapter due to the leading role that for both theories the curvature of space-time in the phenomenon of gravity, we will make a definitive judgment on both conceptions

10 GRAVITY IN THE NEW COSMOLOGICAL MODEL

10.1 The cause of gravity and the meaning of the gravitational constant G

We will start by looking at Figure 25 where we see overlapping the two spheres before and after the creation of the curvature of space-time, which occurs when the mass grows as a curved deformation equal to the arc corresponding to the angle of a radian.

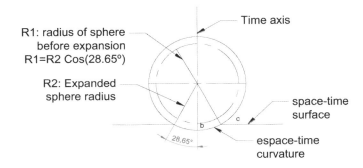

Figure 25 Sphere with the curvature of space-time and internal sphere tangent to the plane of space-time

It is observed that the sphere, expanded after occurring, the curvature of space-time, has a portion of it located below the plane of space-time. According to our model of expansion of the universe, there is an increase in energy with the expansion and consequently with the advance of universal time. To advance in time is to place oneself at a higher energy level, which grows at the speed of light squared. This means that the portion of the sphere in Figure 25 below the plane of space-time will be at an energy level lower than the portion of the sphere that is tangent to the plane of space-time.

Because of this circumstance the points of the sphere that are below the plane of space-time will have a lower energy level the further away they are from the plane of space-time. This means that we could constitute equipotential spheres of negative energy whose value will increase proportional to the distance of their separation from the plane of space-time

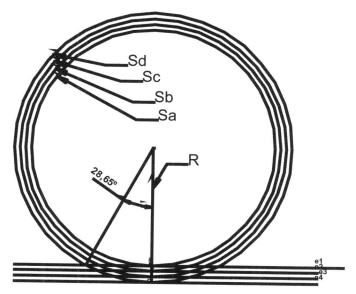

Figure 26 Four concentric spheres tangent to 4 space-time platforms

In Figure 26 we see four concentric spheres Sa, Sb, Sc and Sd that rest at the tangential points a, b, c and d, on four different space-time platforms ETP1, ETP2, ETP3 and ETP4 which have four different energy levels, e1, e2, e3 and e4. Therefore there will be a negative energy differential, between the sphere Sa and the other three spheres, whose value will be directly proportional to the distance at which they are separated. Each of the points of the three spheres will consequently have a negative "PEn" potential energy with respect to the sphere Sa, whose value can be represented as the work "Tn" of a negative force "Fn" multiplied by the distance "Dn" that separates it from the sphere Sa, For the points of b, c, d, with respect to the point "a"

$$PEb = -FbxDab$$
$$PEc = -FcxDac$$
$$PEd = -FdxDad$$

And in general for any point:

$$PEn = -FnxDan \qquad (E10\text{-}1)$$

Since this must be met for all points between the base sphere Sa and the sphere Ed, there must necessarily be a negative vector force field and an associated scalar energy field, which allows the energy differential to be met for all points. As we know the fundamental equation of force is F=m x a therefore what must exist then, is an acceleration field. As the mass corresponding to the negative force is a variable as a function of the radius of the sphere to which the point "n" belongs, we have in consequence that: the negative acceleration field will be associated with a constant value, which will be the mass of the sphere Sa, and with the radius of the sphere containing the point "n". The acceleration field accordingly for a point "n", must correspond to the value of the gravitational field equation of the law of universal gravitation for that point "n":

$$g_n = \frac{G x M}{r_n^2} \quad (E10-2)$$

In this case $M = MSd$

$$g_n = \frac{G x MSd}{r_n^2}$$

As we have mentioned that the equation must be related to the value of the mass of the sphere "a" MSa; if we are right, it will have to be fulfilled that the mass of the sphere "a" MSa will be equal to GxMSd.

$$MSa = G x MSd$$

From where we clear the value of G, the gravitational constant

$$\frac{MSa}{MSd} = G$$

For uniform density:

$$\frac{Volum.Sa}{Volum.Sd} = G$$

$$\frac{\frac{4}{3} x \pi x (Roa)^3}{\frac{4}{3} x \pi x (Rod)^3} = G$$

What: $Roa = Rod\ x\ (cosine\ 28.65º)$

$$\frac{Rod x (cosine\ 28.65º)^3}{Rod} = G$$

$$(cosine\ 28.65º)^3 = G$$

What $cosine\ 28,64785º = 0,87758289$

$$(0,87758289)^3 = 0,675871979 = G \quad (E10-3)$$

It should also be clarified that the constant G has two components:

- **Gm:** The portion of the mass of the body that remains on the plane of space-time, after the curvature of space-time occurs, whose value is Gm = (Cosine 0,5 radian) ³ = 0,675871979

- **Gi:** A value related to the intensity of the gravitational phenomenon, which is very complex to calculate, due to its structure, since it has a part of its mass above and another part below the energy level of the surface of space-time. The experimental value of Gi on Earth, obtained after numerous measurements, is: $G_i = 10^{-10}$ N m² Kg Therefore the value of G will be: $\boldsymbol{G = G_m \times G_i = 6,674 \times 10^{-11} N\ m^2\ Kg^{-2}}$

With obtaining the meaning of the gravitational constant that we have achieved, it is demonstrated without a doubt, that gravity is an energetic phenomenon, which arises as a consequence of the energy gap that occurs with the formation of the curvature of space-time, creating consequently a negative acceleration field, which in the presence of mass produces a vector field of force and a scalar field of energy.

With this discovery of the origin of the phenomenon of gravity, we clarify a secular mystery for humanity and we also obtain further confirmation of the veracity of the foundations of our Theory of the New Cosmological Model, which ratifies the confirmation we previously obtained with Hubble's law. Next we will expand the concepts of the way in which the gravitational phenomenon acts in the universe.

10.2 The gravitational field generated by a spherical body

The gravitational field possessed by spherical bodies is also spherical as shown in the example in Figure 27, and extends to the outside of the body since the space-time platform around the body also has the same negative energy differential with respect to the curvature of space-time.

In Figure 27, the body is a sphere of radius 100 Km, an asteroid. The gravitational field is observed, as equipotential concentric spheres starting from the sphere of the radius of the body in red and the external ones in blue at the distances of, 150 km, 200 km, 250 km and 300 km, from the center of the spherical body.

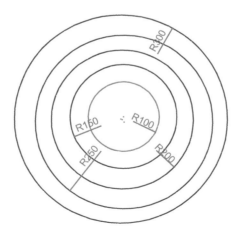

Figure 27 Equipotentials of the gravitational field of a body, generated by its gravity

The gravitational field values for each of the equipotential are calculated using the formula $g = \frac{GM}{r^2}$

In this case we have, considering a density in the body similar to that of the Earth of 5,515 g/cm³:

$$m = \frac{4}{3} \pi . r^3 x 5,515 \, kg/m^3 x 10^{-3}$$

$$m = 2,3113 \, x 10^{12} \, Kg$$

Obtaining the values indicated in Table 7:

Table 7 Gravitational equipotentials

Equipotential	R (Km)	G(Nxm²xKg⁻²)	m (Kg)	g (NxKg⁻¹)
Eq1	100,00	0,675872E-10	2,311E+12	0,493
Eq2	150,00	0,675872E-10	2,311E+12	0,403
Eq3	200,00	0,675872E-10	2,311E+12	0,349
Eq4	250,00	0,675872E-10	2,311E+12	0,312
Eq5	300,00	0,675872E-10	2,311E+19	0,285

We have calculated the acceleration of gravity produced by a body m1 in the environment of it, which in the presence of the mass of another body m2 produces a force of attraction towards the body m1 proportional to the mass of the body m2. We have also calculated the acceleration of gravity at the periphery of the body m1, which multiplied by the mass of the body produces the compressing force of the body m1 on itself.

10.3 Gravity between two spherical bodies

Let's see what happens to the gravitational phenomenon when two spherical bodies approach Figure 28

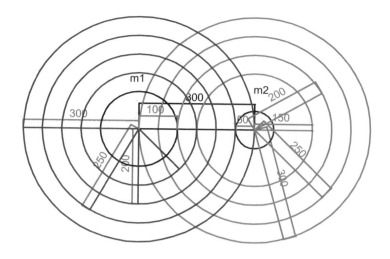

Figure 28 Interaction of the gravitational fields of two spherical bodies

When approaching body m1 to body m2 at a distance R, m1 is introduced into the gravitational field of m2; at the same time, body m2 is introduced into the gravitational field of m1. As a result, the following is generated, all at the same time:

1. Force F1 is created because of the gravitational field effect of m2 acting on the mass of m1.
2. Force F2 is generated because of the gravitational field effect of m1 acting on the mass of m2.
3. A potential energy is induced in body m2 due to the gravitational field of m1.
4. A potential energy is induced in m1 body due to the gravitational field of m2.
5. The energy that receives the smaller m2 body, (example m2 < m1), is transformed into kinetic energy in two ways:
   a)   With an accelerated move towards the center of the larger m1 body causing a collision.
   b)   As a circular motion following the circumference of the equipotential field of m1, which coincides with the center of m2, at the constant velocity resulting from the equation $v = \sqrt{\frac{GM1}{r}}$.
6. The energy received by the body m1 due to the gravitational field of m2 is transformed into kinetic energy by movement following the equipotential curve of the gravitational field

of m2 that passes through the center of m1, at the constant velocity resulting from the equation $v = \sqrt{\frac{GM2}{r}}$.

7. A contractive vector field of force is created on m1 body, and an associated scalar field of energy, due to its own gravitational field g1.
8. A contractive vector field of force is created on m2 body, and an associated energy scalar field, due to its own gravitational field g2.

It should be noted that the forces F1 and F2 are independent and have a different origin, but which is extraordinary is that they are always equal in magnitude and of inverse senses. Although it seems incredible, the force with which the earth attracts the sun, immersed in the gravitational field generated by the Earth, is equal in magnitude and of opposite direction to the force of attraction with which the sun draws to the Earth, immersed in the gravitational field of the sun.

Let's see the demonstration for the example proposed:

$$F1 = m1 x\ g2$$

$$g2 = \frac{Gxm2}{r^2}$$

$$F1 = \frac{Gxm1xm2}{r^2} \quad (E10-4)$$

For F2 we have:

$$F2 = m2 x\ g1$$

$$g1 = \frac{G x \mathbf{m1}}{r^2}$$

$$F2 = M2 x \frac{Gxm1}{r^2}$$

$$F2 = \frac{Gxm1xm2}{r^2} \quad (E10-5)$$

Therefore $F1 = -F2$ (E10-6)

Therefore it is incorrect to say the force of gravitational attraction between bodies M1 and M2, being the right thing to say: the forces of gravitational attraction acting on M1 and M2. The two equal opposing forces act as two people pulling a rope at their ends with equal force. If we include orbital rotation, a simile is that of a hammer throwing athlete on whose rope is equal and in the opposite direction the centrifugal force of the hammer ball and the muscle strength of the athlete retaining the hammer handle. Similarly the energy developed by the athlete retaining the instrument and rotating on his own body, is equal to the kinetic energy generated by circular movement of the hammer ball.

There is other obvious conclusion: gravitational fields of both bodies are not counteract or joined together, on the contrary they coexist without interfering.

With have seen the example of two perfectly spherical bodies in this case, but this is not really what happens in nature, Since most of the bodies are not perfect spheres, due to the rotation on its own axis, are more flattened at the poles and more bulky at its Equator. They are really spheroids.

10.4 Gravity in spheroidal bodies

Bodies in nature are not spherical, since as we mentioned earlier due to the rotational movement, they are more flattened at the poles. This results in them having a larger radius at the equator and a smaller radius in the orthogonal direction to the equator.

If we make a cross-section of the earth by the equator, we will get a circle, and if we make a longitudinal cut by the poles, we will get an ellipse.

This means that the Earth and most of the other bodies of the universe have the form of an ellipsoid of revolution or Oblate spheroid[9].

Let's look at an example of a spheroid in which a=60 and c=40, we will be able to observe the gravitational equipotential on both axes.

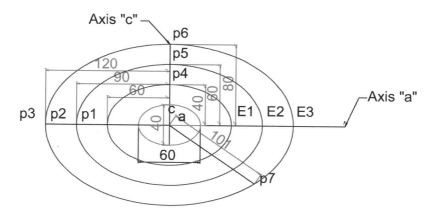

Figure 29 Equipotential gravitational field of a spheroidal body

---

[9] https://en.wikipedia.org/wiki/Spheroid

In Figure 29 on the coordinate axis we can calculate the value of the gravitational field of the equipotentials by applying the formula: $g = \frac{GM}{r^2}$.
Accordingly:

For equipotential E1:

$$g(p1) = \frac{GM}{(60)^2} = \frac{GM}{3600}$$

$$g(p4) = \frac{GM}{(40)^2} = \frac{GM}{1600}$$

For equipotential E2:

$$g(p2) = \frac{GM}{(90)^2} = \frac{GM}{8100}$$

$$g(p5) = \frac{GM}{(60)^2} = \frac{GM}{3600}$$

Para la equipotencial E3:

$$g(p3) = \frac{GM}{(120)^2} = \frac{GM}{14400}$$

$$g(p6) = \frac{GM}{(80)^2} = \frac{GM}{6400}$$

We get on two different values for the same equipotential, which is absurd. What requires analysis:

What happens is that in spheroidal bodies we no longer have a sphere as the creator of the curvature of space-time, but rather a spheroid, which has a variable radius of curvature and consequently a variable vector radius, which is why the gravitational constant will also have to be variable depending on the radius of curvature.

Let's define as Gc, the corrected value of G, which will necessarily have the following equation:

$$Gc = \frac{G x Rc^2}{a^2} \quad (E10-8)$$

Rc being the radius vector and "a" being the value of the semi-axis "a" of the corresponding equipotential. Therefore:

$$Gc1 = \frac{Gx(60)^2}{(60)^2} = Gc2 = \frac{Gx(90)^2}{(90)^2} = Gc3 = \frac{Gx(120)^2}{(120)^2} = G$$

$$Gc4 = G\frac{(40)^2}{(60)^2} = Gc5 = G\frac{(60)^2}{(90)^2} = Gc6 = G\frac{(80)^2}{(120)^2} = 0{,}44\,G$$

$$Gc7 = G\frac{(101)^2}{(120)^2} = 0{,}708\,G$$

We define: $K = \frac{Rc^2}{a^2}$, , being able to express the formula of Gc, as follows

$$Gc = K(\theta)G \quad (E10-9)$$

As K, is a function of the radius vector and the vector radio changes with the angle to the shaft "a", in consequence the function K, we can express it as an angular function K(θ), being θ the angle about axis 'a', K(θ) has a value at least for θ = 90° and a maximum value r = 1 to θ = 0°

As a result we then formulate the following theory:

10.4.1   Theory the Spheroids Gravitational Fields

The equipotential gravitational fields of the Oblates spheroids bodies are equally Oblates spheroids. In equipotential said the value of G to make it ceases to be constant in the "c" axis of symmetry, by taking a value defined by the equation Gc=KG , K is a variable that is defined by the equation K=Rc²/a², where Rc is the radius vector and "a" the value of the axis "a" of the corresponding equipotential.

10.5 Gravity between two spheroidal bodies

We have already seen how operates the gravity between two spherical bodies, but that is a theoretical exercise since bodies are not truly spherical. As bodies in nature are spheroids shape as it operates the gravity between them because it is not a theoretical case, but real and that reality has been known since the year 1609. In that year Johannes Kepler, formulated the first of the empirical laws that bear his name, arising from the interpretation of astronomical observations carried out carefully by his Professor Tycho Brahe.

Kepler's laws describe accurately the movement of the planets in the solar system, their validity is not in discussion. However, there is no conclusive answer to the following questions:

1) Because the orbits are elliptical?
2) Because the Sun is one of the focus of the ellipse and not in the Center?
3) That determines the eccentricity of elliptical orbits?
4) Because the planets have a tilt on its axis of rotation in the plane of the ecliptic?
5) Because the inclination of the axis of rotation of the Earth, specifically, is 23.5 ° on the ecliptic plane?

These questions have been unanswered for 400 years, it seems difficult to find a scientific explanation for these questions after so long without anyone having previously clarified them. But we'll try……

The first thing we need to consider are two relevant and determining facts for the study:

1) Even though the bodies and their gravitational equipotential are three-dimensional, the gravitational phenomenon of a stable orbit materializes in two dimensions, on a specific plane called an ecliptic.
2) The gravitational phenomenon of a stable orbit shall occur only between two bodies on their ecliptics. This means that if a body is orbited by multiple bodies, there will be a separate ecliptic plane for each case.

We must therefore look as both bodies and their respective gravitational equipotential are interrelated on the plane of the ecliptic. The branch of geometry that studies the representation of a three-dimensional body, on two dimensions on a plane, is called descriptive geometry. As for the study of this case it is necessary to use descriptive geometry[10].

When it intersects or sections a solid with a plane you get different geometric figures on the level of court, depending on the angle of the plane with respect to the axis of the solid. The example most known and widely studied from the Greek civilization, are Conic sections, if we intersect a cone with a plane, obtain depending on the angle of intersection: 1) a circle; (2) an ellipse; (3) a parabola; (4) a Hyperbola

In the event that we are concerned it is the intersection, in the plane of the ecliptic, of two spheroids, the rotated body (m1) and (m2) Rotator body. The intersection of a spheroid by a plane, the plane produces an ellipse, which will vary in their dimensions and eccentricity according to the angle of intersection.

Remember, as mentioned above, that the gravitational attraction is not a force that acts on two bodies (m1 and m2). They are two independent forces equal in magnitude but opposite direction.

$$F1 = m1 \mathrm{x} g2 = \frac{m1 \mathrm{x} m2 \mathrm{x} G}{r^2}$$

$$F2 = m2 \mathrm{x} g1 = \frac{m2 \mathrm{x} m1 \mathrm{x} G}{r^2}$$

$$F1 = -F2$$

It follows from the formulas, that the only way that both forces are equal in magnitude it is necessary that the value of G is the same in both equipotential, i.e. the value of G at the point of the equipotential of g2 which passes through the center of m1 should have the same the value of G,

---

[10] https://en.wikipedia.org/wiki/Descriptive_geometry

at the point of the equipotential of g1 that passes through the center of m2.

The plane of the ecliptic intersects the bodies m1 and m2 in the half of the coordinate Z axis. In the case of spherical bodies whatever the angle of inclination of the axis of rotation, the intersection of the ecliptic plane will result in a circumference of diameter equal to that of the sphere, being constant the value of G equal to that of the gravitational constant or is $G = (COSINE\ 0.5\ rad)^3$. Therefore, F1=F2 will always be met. And the orbit will be circular over one of the circular equipotential of the m1 rotated body.

In the case of the bodies are spheroids, also the flat ecliptic intersects the bodies m1 and m2 in the half of the coordinate Z axis, which corresponds in the ellipsoid with its axis of symmetry. But there is a notable difference in this case and is the result of the intersection will vary from according to the degree of inclination of the axis of rotation. If the axis of rotation of both bodies don't have no inclination, in this case the intersection will be a circle and the equipotential are circular with a constant value of Gc = 1 G and rotation will be the same as in the case of spherical bodies, around a circular equipotential of the rotated body m1. But if any of the bodies has tilted its axis of rotation their intersection is an ellipse with a variable value of the gravitational constant. The ellipse of the interception will be one major axis equal to the more "a" of the spheroid intersected axis and an axis minor b, variable according to the angle of the intersection.

In this case necessarily materialize the gravitational effect in a stable orbit must be met for the following conditions:
1) The resulting ellipses from the intersection of both bodies in the plane of the ecliptic spheroids must have the same eccentricity
2) The minimum value of Gc coincident with the axis of the ellipse b should be equal in both cases.
3) The value of Gc coincident with the major axis of the ellipse must be equal to 1G.
4) For any position of the orbit both bodies should maintain parallel their axes of coordinates.

The third condition always met, either that is the inclination angle of the spheroids with respect to the ecliptic plane as it intersects. Since the value of the major axis of the resulting ellipse of the intersection will always be equal to the value of the 'a' of the ellipsoid axis and we know that in this case its value is 1 G.

To achieve conditions 1) and 2), is necessary that the angle of inclination of each spheroid on the ecliptic plane as it intersects both cases corresponds to the same value of Gc. The value of Gc on a spheroid varies in the C-axis of symmetry according to the angle of the vector r between a maximum value of 1 G to zero degrees and a minimum value of Gc to 90 °, where the different minimum value according to the eccentricity of the spheroid. In Figure 30, we can observe how the Gc value depending on the angle of the radius vector with respect to the axis orthogonal to the axis of rotation Z

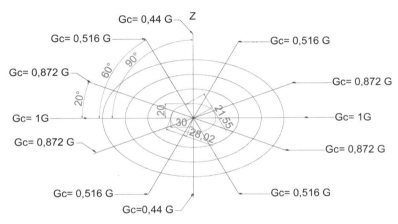

Figure 30 Value of the gravitational constant Gc, according to the angle of the vector radius with respect to the "a" axis

In the upper left quadrant we have the radio vectors at 20°, 60° and 90°, with respect to the axis orthogonal to the axis of rotation Z, of an ellipsoid of dimensions a = 60 and C = 40. The gc values are obtained by applying the formula (E10-8), designed in the previous chapter:

$$Gc\,(20º) = K(20º)G = \frac{(28,02)^2}{(30)^2}G = 0,872G$$

$$Gc\,(60º) = K(60º)G = \frac{(21,55)^2}{(30)^2}G = 0,516G$$

$$Gc\,(90º) = K(90º)G = \frac{(20)^2}{(30)^2}G = 0,44\,G$$

The values obtained from Gc will be repeated in each equipotential with the same angle of the vector radius and equivalently in the other three quadrants, as shown in Figure 30.

For the above we can achieve, that two spheroids of different eccentricity produce ellipses paths of equal symmetry if we intersect them with the ecliptic plane by the respective angles of inclination corresponding to the same value of Gc.

We can explain this better with an example:

Suppose a spheroid M1 of mass of high density and dimensions a=60 and C=40, must be orbited by another spheroid M2 of mass of lower density and of dimensions a = 70 and c = 20. M1 has an inclination over the ecliptic plane of 51° as shown in Figure 31

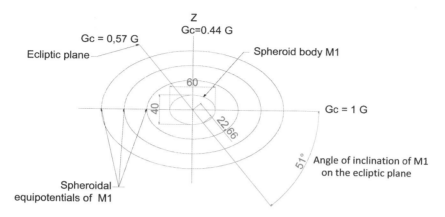

Figure 31 Intersection of the body M1, by the ecliptic plane, with an inclination of its axis of rotation of 51°

Gc values: Gc (90°) =K(90°)G= (20)²/(30)²G=0.44 G; Gc (51°)=K(51°)G= (22.66)²/(30)²G= 0.57 G.

In Figure 32, we see the image of the intersection recorded in the plane of the ecliptic

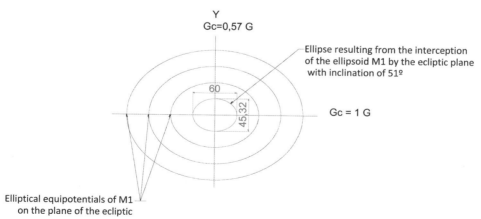

Figure 32 Ellipse and elliptical equipotentials, obtained in the ecliptic plane by the intersection of that plane on M1 with an inclination of 51° on its axis of rotation

To meet the necessary conditions for a stable orbit the inclination of M2 over the ecliptic plane must correspond to the angle to which it corresponds to the same value of Gc obtained at the intersection of M1, i.e. Gc = 0.57 G. For which we proceed as follows:

The spheroid M2 is rebutted on the "a" axis, as if it were a revolving door, until we obtain the corresponding value of Gc, in this case Gc = 0.57G. This value resulted in an angle of 15°, as shown in Figure 33. For such an angle the modulus of the radius vector is 26.4, therefore:

$$Gc(15º) = K(15º)G = \frac{(26,4)^2}{(35)^2}G = 0,57G$$

$$Gc(15º) = 0,57G$$

For Gc(90°)

$$Gc(90º) = K(90º)G = (20)^2/(70)^2\,G = 0,0816\,G$$

$$Gc(90º) = 0,0816\,G$$

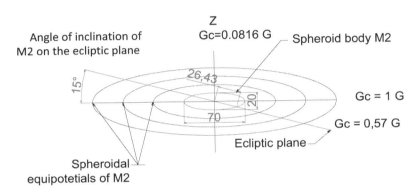

Figure 33 Ellipse and elliptic equipotentials, obtained in the ecliptic plane by the intersection of that plane on M2 with an inclination of 15° on its axis of rotation

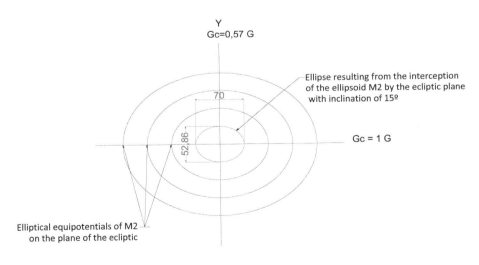

Figure 34 Ellipse and elliptic equipotentials, obtained in the ecliptic plane by the intersection of that plane on M2 with an inclination of 15° on its axis of rotation

Next we will verify that the ellipses resulting from the intersection of both spheroids by the plane of the ecliptic have the same eccentricity.

We will proceed to calculate the eccentricities of both ellipses:
Ellipse M1: a=30; b=22.66

$$c = \sqrt{a^2 - b^2} = \sqrt{30^2 - 22.66^2}$$
$$c = 19.66$$
$$\varepsilon = \frac{c}{a} = \frac{19,66}{30} = 0,655$$

Ellipse M2: a=35; b=26,43

$$c = \sqrt{35^2 - 26,43^2}$$
$$c = 22,94$$

$$\varepsilon = \frac{22.94}{35} = 0,655$$

Well, the first three conditions necessary to generate a stable orbit have been met. It is only necessary to simulate the orbit by keeping the coordinate axes of both bodies parallel throughout the trajectory of the orbit, in order that the angle of the radius vector with respect to the axis "a" is equal in both bodies M1 and M2 and therefore have the same value of Gc and it can be fulfilled that always the gravitational attraction force F1 is equal and contrary to the force gravitational F2, i.e. F1=-F2. We will simulate the orbit of an M2 planet, on an M1 star in a planetary system similar to the solar system.

We will start by locating M2 in the periaster, for which we will place M2 at a distance D1 x 200 million Km from the center of M1, as shown in Figure 35

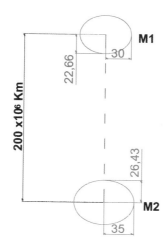

Figure 35 Bodies M1 and M2 at the distance of the perister

By setting the distance of the periaster, we only need to find out the eccentricity of the orbit, since with only those two data the properties of the geometric figure called ellipse allow us to completely define the shape of the orbit. If successful, we would get a surprising result and it is the fact that the shape of the planetary orbits is independent of the mass of the bodies. We'll see if that's possible, at least we'll try.

There is an equation that relates the periapsis distance "q" with the value of the more "a" axis and "ε" the ellipse of the orbit eccentricity. Which given below:

$$q = a(1 - \varepsilon) \qquad (E10\text{-}10)$$

The "q" value varies between 0 and 1, achieving its maximum value when ε = 0, which corresponds to an orbit circular characteristic of a spherical body whose intersection with the ecliptic plane is a circle. We accordingly conclude that the value of ε varies depending on the eccentricity of the ellipse that is registered in the interception of the rotated body M1 in the plane of the ecliptic.

On the other hand "q" is the distance between M1 and M2 and corresponds to the dimensions of the semi-minor axis the equipotential of M1 that passes through the center of M2. The equipotential have the same eccentricity, the ellipse that is registered in the interception of the rotated body M1 in the plane of the ecliptic and the eccentricity will vary according to the value of Gc and Gc = KG, the distance "q" will vary depending on K (90 °). For K (90 °) = 1, the ellipse is a circle, which demonstrates the relationship between both variables K (90 °) and ε.

To find the equation that relates to both variables, will resort to the equivalence principle a famous artifice, considered one of the greatest geniuses of Albert Einstein, the famous trick of the elevator, which allowed him to demonstrate that an observer locked inside an elevator there would be media perceive if it were falling due to gravity or were floating in an inertial State.

In our case the trick would be the following:

In elliptical orbits only in the periapsis and the apsis, because the trajectory is orthogonal to the vector radius, we can calculate the tangential velocity of the M2 rotator body. Consequently in the periapsis, we can calculate the tangential velocity using the formula:

$$1) \quad v = \sqrt{\frac{(K(90º)G\, M1)}{q}} \qquad (E10-11)$$

What corresponds to the case of an ellipsoid, however that same velocity is obtained from a spherical body of the same mass with a greater vector radius, in this case the vector radius will be R=q/K(90°), therefore, therefore:

$$2)\ v = \sqrt{\frac{(G\ M1)}{q/K(90º)}} \quad (E10-12)$$

Speeds are the same, but there is a big difference, in one case the speed corresponds to a spheroid distance q and the second to a sphere of the same mass with an orbit of ε =0 and the radius vector equal to the value of the semimajor axis "a". So we can write the equation (E10-12)

$$3)\ v = \sqrt{\frac{(G\ M1)}{a}} \quad (E10-13)$$

An observer located on M2, does not like knowing if its speed of displacement v is the result of the equation (E10-10) or equation (E10-11) or the equation (E10-12) therefore they are physically equivalent.

As a result the following seems obvious:

$$a = \frac{q}{K(90º)}$$

As we knew before:

$$q = a(1 - \varepsilon)$$
$$a = \frac{q}{1 - \varepsilon}$$

$K(90º) = 1 - \varepsilon$

Finally:

Equation $(E10-13)$ *of the eccentricity of planetary orbits*

$$\varepsilon = 1 - K(90º) \quad (E10-14)$$

This result has momentous consequences, as we have scientifically shown that the elliptical orbits of the planets around the Sun and all orbits in general, owe their shape to the geometric characteristics of the bodies and not to their masses

After the end of the chapter we will make a set of conclusions and new theories resulting from this discovery. We will continue so soon, with the calculation of the orbit of the example and subsequently, with the calculation of the orbit of a planet of the solar system as a test of this theory

For the case of the example:

Distance of the periaster= $200x10^6 Km$; K(90°) = 0.57

So that:

$$a = \frac{q}{K(90º)} = \frac{200}{0,57} = 350,88$$
$$a = 350,88\ x10^6 Km$$
$$c = a - q = 350.88 - 200 = 150,88$$

$$c = 150{,}88 \; x10^6 Km$$
$$b = \sqrt{a^2 - c^2}$$
$$b = \sqrt{350{,}88^2 - 150{,}88^2} = 316{,}78$$
$$b = 316{,}78 \; x10^6 Km$$

Distance of the apoastro Q:
$$Q = a + c = 350{,}88 + 150{,}88 = 501{,}76$$
$$Q = 501{,}76 \; x \; 10^6 Km$$

With the data obtained we will proceed to draw the orbit:

In Figure 36, we have the elliptical orbit, with 6 positions of M2 in the orbit.

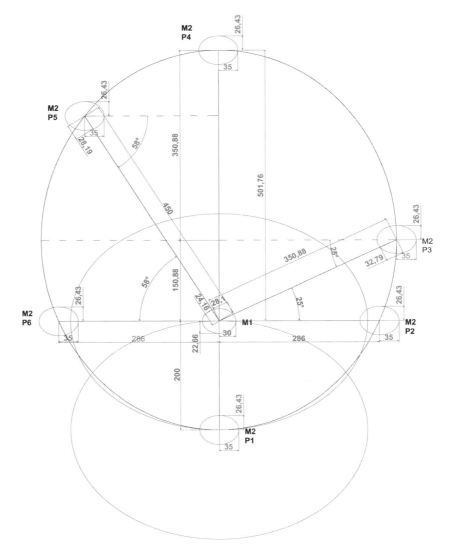

Figure 36 Orbit between two spheroidal bodies. Scale 1:1x10^6

In Figure 36, we have drawn only 2 equipotentials so as not to reload the Illustration

We will proceed in the first instance to check that in each of the 6 points of the orbit, it is true that F1 =-F2.

For point No. 1 of the orbit:

$$F12 = M1 x g2 = M1 \times M2 \frac{Gc_{21}}{r_{21}^2}$$

$$F21 = M2 x g1 = M2 \times M1 \frac{Gc_{12}}{r_{12}^2}$$

For this case $Gc_{12}=Gc_{21}$ =K(90°)G = 0.57 G y r_{12} =200, angle (-90°); r_{21}=200, angle 90°

$$F12 = M1xM2 \frac{0{,}57G}{(200)^2} = -F21 = M2xM1 \frac{0{,}57G}{(200)^2} = -M2xM1xG \; x14{,}25x10^{-6}$$

For this points 2 y 6, $Gc_{12}=Gc_{21}$ = K(0°)G = 1G y $r_{12}=r_{21}$ = 286, angle 0°

$$F12 = M1xM2 \frac{G}{(286)^2} = -F21 = M2xM1 \frac{G}{(286)^2} = -M2xM1xG \; x12{,}22x10^{-6}$$

For point 3: r_{12} =350.88, angle 25°; r_{21}=350,88 angle (-25°)

$Gc_{12} = K_{12}(25º)G = G(28.1)^2 /(30)^2 = 0{,}877 \, G$

$Gc_{21} = K_{21}(-25º)G = G(32{,}79)^2 /(35)^2 = 0{,}877 \, G$

$$F12 = M1xM2 \frac{0{,}877\,G}{(350{,}88)^2} = -F21 = M2xM1 \frac{0{,}877\,G}{(350{,}88)^2} = -M2xM1xG \; x \; 7.12 \, x10^{-6}$$

For poit 4; $Gc_{12}=Gc_{21}$ = K(90°)G= 0.57 G y r_{12} =501,76, angle (90°); r_{21} =501,76 angle (-90°)

$$F12 = M1xM2 \frac{0{,}57G}{(501{,}76)^2} = -F21 = M2xM1 \frac{0{,}57G}{(501{,}76)^2} = -M2xM1xG \; x \; 2{,}26x10^{-6}$$

For poit 5; r_{12} =450 angle 58°; r_{21}=450 angle (-58°)

$Gc_{12} = K_{12}(58º)G = G(24{,}16)^2 /(30)^2 = 0{,}648 \, G$

$Gc_{21} = K_{21}(-58º)G = G(28{,}19)^2 /(35)^2 = 0{,}648 \, G$

$$F12 = M1xM2 \frac{0{,}648G}{(450)^2} = -F21 = M2xM1 \frac{0{,}648G}{(450)^2} = -M2xM1xG \; x \; 3{,}2x10^{-6}$$

10.5.1 Analysis of the results of the gravity experiment between two spheroidal bodies

I. It is verified that at the 6 points of the orbit it is met that F1-F2

II. It is verified that the 4 conditions initially raised are essential for the establishment of a stable orbit, i.e.:

   a) The ellipses resulting from the intersection of the two spheroidic bodies, by the plane of the ecliptic, must have the same eccentricity.

b) The minimum value of Gc matching the b-axis of the ellipse must be the same for both cases.

c) The Gc value that matches the major axis of the ellipses must be equal to 1G.

d) For any position in the orbit both bodies must keep their coordinate axes parallel

III. The experiment accurately reproduces the 4 seasons that present Earth, Mars and other planets:

Solstices:
- M2 (planet) at position P1, of the periaster, has an inclination of 15o over the vector radius of M1 (star) which corresponds to the summer in the Northern Hemisphere and winter in the Southern Hemisphere.
- M2 at the P4 position of the apoastro, has an inclination of (-15°) over the vector radius of M1, which corresponds to summer in the southern hemisphere and winter in the Northern Hemisphere.

Equinoxes
- M2 at position P2, has the inclination of 15° in the orthogonal plane to the vector radius of M1, which corresponds to autumn in the Northern Hemisphere and spring in the Southern Hemisphere
- M2 at position P6, has the inclination of 15° in the orthogonal plane to the vector radius of M1, which corresponds to autumn in the southern hemisphere and spring in the northern hemisphere.

But the most important and surprising thing about this experiment and that it is an extraordinary and unprecedented revelation is that the shape of the elliptical orbit is obtained from the geometry of the bodies and their inclinations on the plane of the ecliptic regardless of the value of their masses.

As the experiment was done on a hypothetical case, to validate this important revelation, we must check whether it is possible to obtain the orbit of a planet from the solar system, starting only from its dimensions and the angle of inclination of its axis of rotation. We will choose Saturn's planet whose spheroid has an appreciable eccentricity. The other planets in the solar system have very little eccentricity, so the results of an experiment would not be reliable, because the margin of error of the data is similar to the margin of error of the calculations:

10.5.2 Calculation of the orbit of the planet Saturn around the Sun, based only on the data of its geometry and the inclination of its axis of rotation.

We previously consulted the most relevant data of the planet Saturn in the free encyclopedia Wikipedia:

Equatorial diameter: 120.536 km.

Polar diameter: 108.728 km

Axial inclination: 26,73°

periaster or Perihelio: q= 9,04807635 UA

We will proceed to the calculation of the ellipse that is recorded in the plane of the ecliptic with an inclination of Saturn's axis of rotation of 26.73°

In Figure 37, we see Saturn's spheroid intersected by the plane of the ecliptic with an angle of 26,73°.

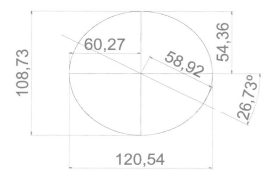

Figure 37 Saturn's spheroid intersected by the ecliptic plane, at an angle of 26.73° from its axis of rotation; scale 1:1.000 Km

In Figure 38, we see the ellipse that is recorded after the intersection in the plane of the ecliptic.

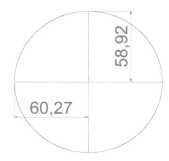

Figure 38 Ellipse that is recorded in the ecliptic plane after the intersection; scale 1:1000 Km

We calculate the value of Gc (90°) = K(90°)G, using equation (E10-8):

$$K(90º) = \frac{(58,92)^2}{(60,27)^2} = 0,955$$

$$K(90º) = 0,955$$
$$Gc(90º) = 0,955\ G$$

We know: $q = a(1 - \varepsilon)$,
$$a = \frac{q}{1 - \varepsilon}$$

And as we demonstrated earlier $K(90º) = (1 - \varepsilon)$
$$a = \frac{q}{K(90º)} = \frac{9,048}{0.955} = 9,474$$
$$a = 9,474\ UA$$

Using the properties of the ellipse, we calculate the rest of the values of the elliptical orbit.
$$c = a - q = 9,474 - 9,048 = 0,426$$
$$c = 0,426\ UA$$
$$b = \sqrt{a^2 - c^2} = \sqrt{9,474^2 - 0,426^2} = 9,464$$
$$b = 9,464\ UA$$

Finally the apoastro:
$$Q = a + c = 9,474 + 0.426 = 9,90$$
$$Q = 9,90\ UA$$

From the data provided in the Wikipedia reference, we calculate the values of "c" and "b" of the actual orbit:
$$c = a - q = 9,537 - 9.048 = 0,489\ UA$$
$$c = 0,489\ UA$$
$$b = \sqrt{a^2 - c^2} = \sqrt{9,537^2 - 489^2} = 9,524\ UA$$
$$b = 9,524\ UA$$

Below is the comparative table 8 of the actual and calculated values of Saturn's orbit:

Table 8 Saturn's actual orbit and calculated orbit

Parameter	Symbol	Units	Real orbit	Calculaed orbit	Deviation	Percentge deviation	Percentge accuracy
periaster	q	UA	9,048	9,048	X	X	X
Medium radius	a	UA	9,537	9,474	0,063	0,66%	99,34%
Apoastro	Q	UA	10,115	9,9	0,215	2,13%	97,87%
Minor semiaxis	b	UA	9,524	9,464	0,06	0,63%	99,37%

The results are surprisingly almost exact to the real

In Figure 39 we draw superimposed the two orbits. They're two almost identical orbits

It is rewarding when in the development of research, a real and blunt proof of a theory previously demonstrated mathematically is achieved and fortunately that happens in this case

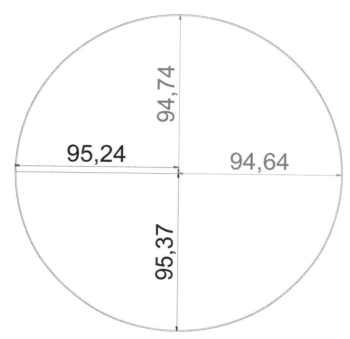

Figure 39 Orbits of Saturn: in black real orbit; in calculated orbit red;scale 1: 0.1UA

As a result of this strong demonstration we can at all good property take full, after 400 years of the first law of Kepler, the five questions of the beginning of the chapter:

1) Because the orbits are elliptical?

2) Because the Sun is located in one of the spotlights of the ellipse and not in the center?

3) What determines the eccentricity of elliptical orbits?

4) Why do planets have an inclination on their axis rotation over the plane of the ecliptic?

5) Because specifically the inclination of the earth's axis of rotation is 23.5° on the ecliptic plane?

In conclusion of what is stated in this chapter, we formulate the following new theories:

10.5.3 Theory of "The Cause of the Inclination of the Axis of Rotation of Earth and Other Planets"

"The inclination of 23.5° of the Earth's axis of rotation is due to the Earth's need to form an ellipse at the intersection with the plane of the ecliptic, of equal eccentricity, as the ellipse that forms the Sun at the intersection by the plane of the ecliptic , on which it has an inclination of 7°. The four stations originate due to the need of Earth throughout the orbit path to permanently parallelize the coordinate axes of the ellipse that generates its intersection with the plane of the ecliptic with the coordinate axes of the ellipse that is generated with the intersection of the Sun with the plane of the ecliptic, so that the gravitational pull force of the earth on the Sun, is equal and counter-clockwise to the gravitational pull force of the Sun on Earth. The same thing similarly happens with the other planets"

10.5.4 Theory of "The cause of the elliptical form of planetary orbits defined in Kepler's first law"

"The elliptical shape of the orbits of the planets around the Sun and the other gravitational orbits of the universe, as established by Kepler's first law, is determined exclusively by the geometry of the bodies regardless of the value of their masses. Obviously the magnitude of the orbits is directly related to the value of their masses, but their shape is due exclusively to the geometry of the bodies and the angle by which their axis of rotation are intercepted by the plane of the ecliptic. The eccentricity value of the elliptical orbit is given by the equation $\varepsilon = 1 - K(90º)$, being $K(90º)$ the value of the angular function K(Θ) of adjusting the gravitational constant to 90° relative to the "a" axis of the ellipses that are recorded in the interception that the plane of the ecliptic performs on the rotation axes of the 2 bodies of the orbit. The meaning of the angular function K(Θ) is defined in the Theory of "The Spheroidal Gravitational Fields""

Finally there is an important conclusion: The scientific explanation obtained from Kepler's First Law in this chapter is a great confirmation by practical physics to the Theory of the New Cosmological Model by the multiple observations over 400 years that support Kepler's laws and joins the confirmations previously obtained, in the previous chapters, by the scientific explanation of Hubble's law and The Law of Universal Gravitation.

10.6 Time is not changed by the effect of gravity

Time can never deform because as we said before it is unique throughout the universe, it only changes the perception of an observer's time ac-cording to their position. An observer floating in space away from mas-sive bodies will observe that time passes exactly the same as the time of the universe, in-stead, an observer located on a massive body, will ob-serve a delay in time proportional to the depth of the curvature of the space-time, which is caused by its implantation on that surface, since the depth of the curvature of time space, is on the Z-hyperspace axis parallel to the axis of time. This means that the perceived delay of time will be directly proportional to the gravity of the body.

So where the perception of time within a massive astro, logically ends until its gravitational field is significant, that is, the frontier of the environment corresponding to the time of the universe and that of the time of a massive body is diffuse and the distortion in the perceived time by an observer will decrease proportionally to their estrangement from the massive body. The real-time of the universe will continue to march fast and fast at the speed of light at which the plane of space-time moves within each Miniverse.

There is no change in time by gravity, what happens as we have said before, is that gravity marks the frontier of the body's time perception realm that generates gravity.

10.7 Conclusions on the origin of gravity in the New Cosmological Model

With this discovery of the origin of gravity and the exact magnitude of the gravitational constant, we clarified a secular mystery for humanity and a series of doubts and controversies for the scientific world, since in the year 1687 when Isaac Newton published his theory of universal gravitation. In our view, the main doubts and controversy have been the following:

1. Is gravity a phenomenon that owes its origin to the structure of matter and therefore must be explained by quantum mechanics, or is it a phenomenon of external origin explained by classical physics?

2. Gravitational pull is an instantaneous force that is created when two bodies are placed in front of each other as exposed in his New-ton theory, or on the contrary, as Einstein says in his theory of general relativity, there cannot be an instant force because noth-ing can travel faster than the speed of light?

The answer to the first question is obvious, the phenomenon of gravity as we have shown in this research work corresponds to a cause external to matter and therefore its explanation responds to classical physics.

Before answering the second question, we will review the concept that gravity is just a geometric phenomenon as established by Einstein's General Theory of Relativity

Einstein's idea is that a massive body will produce a curved distortion of the fabric of space-time, thus creating a concave depression propor-tional to the mass of the body so that any lower mass body would tend to precipitate towards the body simply for geometric reasons and not because of the grav-itational pull force.

Wikipedia next Figure is observed as a massive body distorts space-time, producing a curvature in the form of a concave cavity.

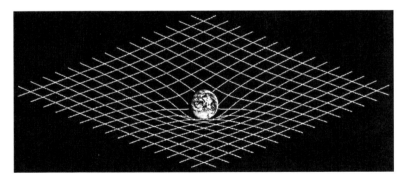

Wikipedia's picture..

Below we raise our main objections to this theory.

1. In case a body of lower mass m2, is located in the vicinity of the cavity where at the bottom of it is housed the larger body M1, that body m2 will also create and be located at the bottom of a smaller concavity. That forces m2 to exit its cavity and make a movement in the direction of the M1 cavity?

2. In the course, inexplicably, it would come out of its cavity and move to the edge the cavity of M1 and roll down the slope in free fall. What would prevent him from ending his career colliding with M1?.

3. If that were the previous case. How can there be a collision between two bodies without a force?

4. If instead of colliding it created, out of nowhere, an orbital rotational movement of the m2 body around the M1 body. What energy would permanently maintain that movement?. The known case of a roulette bowl, in a casino, which spins by the impulse printed by the "croupier", ends its movement irretrievably by consuming the energy of the impulse.

5. If instead of a body were several bodies that would "fall" into the pit of M1, example m2, m3 and m4. What is shown in Figure 40 will happen. In the case of the example, none of the planes (ecliptic) of the orbits of the three bodies pass through the center of the orbited body and M1 in none of the cases is found in one of the focus of an elliptical orbit. This situation is incompatible with the laws of Kepler and the reality observed in the universe.

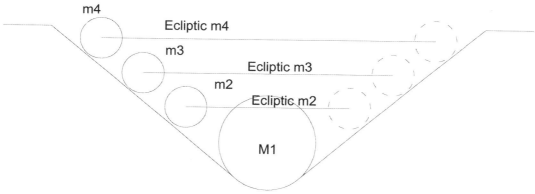

Figure 40 Ecliptics in the General Theory of Relativity

6. How can it be explained according to this theory, the shape of the solar system, which has almost on the same plane the 8 ecliptic that forms the Sun with its 8 planets, without forming a concave pot?

We can conclude on the basis of the 6 arguments cited above and the other concepts set out in this Chapter 10, without a doubt, that it is ruled out that gravity is a geometric phenomenon, due to the deformation of space-time according to the model of the curvature of space-time of Einstein's Theory of General Relativity

After this analysis, the answer to the second question is clear: **Newton was right**; it's not a geometric problem and the space being curved. Gravity occurs due to an energy imbalance, caused by the curvature of time-space according to the Theory of the New Cosmological Model, which generates a gravitational field that in the presence of a body induces a force and energy proportional to the mass of the body.

While it is true that any field, such as the electromagnetic field, needs a finite time to expand, in the case of gravity, that field travels already expanded with matter, so that by encountering two bodies that each carry their field ahead of them gravitational, both fields connect and interact when they come into contact almost instantly so you don't have to mediate the time span that elapses for light to arrive from one body to another to establish gravitational pull.

But other conclusions about the gravity of the investigation:

1) Gravitational pulling takes place between two bodies in two dimensions on a common ecliptic plane without interference from other bodies.
2) One body can be gravitationally related to more than one body, but in orbits of independent ecliptic planes without interference from other bodies.
3) The equation $v = \sqrt{GM/r}$ is only valid for the theoretical case of a spherical body orbited by another spherical body. Therefore its usual use to calculate the approximate mass of galaxies is totally wrong
4) There is and cannot be an equation that gravitationally relates, through the law of universal gravitation, to a set of bodies greater than 2.

The above 4 points are fully confirmed by calculation made in this investigation of the immense orbit of the planet Saturn with an accuracy of 99.34%, based only on the eccentricity of its body and the angle of inclination of its axis of rotation. This orbit only depends exclusively on the Sun and Saturn, without interference from the orbits that make up Saturn's 62 satellites and the orbits of the other seven planets and other bodies (asteroids, comets, etc.) Orbiting the Sun

10.8 The model of the curvature of space-time of the Theory of the New Cosmological Model is confirmed

In chapter 9 we raise the differences between the time curvature model of space proposed by Einstein's Theory of General Relativity and the model proposed for the same purpose by the Theory of the New Cosmological Model. Considering the results and conclusions of this investigation in relation to the gravitational phenomenon, we conclude the following:

1) The model of the curvature of space-time proposed in Einstein's Theory of General Relativity is discarded, based on the following arguments:
   a. As shown in Figure 40, this model is inconsistent with Kepler's laws which are widely confirmed by countless astronomical observations made around the world for 400 years.
   b. This model is incompatible with the expansion of the universe, since in the expansion model proposed by the Theory of the New Cosmological Model, proven by Hubble's law, cannot occur a curved distortion of time-space in a different area than the direct contact between the spherical body that produces the curvature and the plane of space-time.
2) It confirms the model of curvature of space-time proposed by the Theory of the New Cosmological Model for the following reasons:
   a. The model is confirmed by the discoveries of the cause of gravity and the exact value of the gravitational constant G discovered in this research based on that space-time curvature model.
   b. The model is not only fully compatible with Kepler's laws, it is also part of the gravitational model of spheroidal bodies that scientifically demonstrate Kepler's first Law as established by the Fifteenth Theory of this research.
   c. The model is fully compatible with the expansion of the universe as its formulation arises precisely as a result of the expansion of the universe.

Consequently, the theory of this research is confirmed which states that all celestial bodies form a spherical cap on the surface of space-time according to the dimensions stipulated in Chapter 9.1 of this research. Confirmation of the spherical cap model that forms the curvature of time space as set forth in Chapter 9.1 is of paramount importance for the study of stellar black holes. Study that we will address in future chapter.

10.9 Theory of modification of the law of universal gravitation

As a result of this investigation, the Universal Gravitation Act is amended with the following changes:

a. **Gravity is caused by an energy differential between parts of a body, as a result of the curvature of space-time, which results in part of the body acquiring a negative potential energy, relative to the part of the body that remains about the space platform time above the curvature.**

b. **The exact value of the universal gravitation constant is equal to:** $G = \left(cosine\,(0.5\,radian)\right)^3 x\,10^{-10}\,N.m^2/Kg$. **This theoretical value corresponds to a perfectly spherical body and can be particularized for each body according to its geometry.**

Finally, for all that is widely demonstrated in this chapter:

10.10 Einstein's theory of General Relativity is discarded, as it is incompatible with the expansion of the universe and Kepler's First Law.

10.11 Light is not curved by the effect of gravity

The belief that light curves by the effect of gravity has its origin in the astronomical experiment carried out by Arthur Eddington, in 1919, in which he was able to observe during a solar eclipse how light from distant stars curved as he passed near the Sun. This effect was attributed by Albert Einstein specifically to the curvature of space time conceived in his General Theory of Relativity only as something geometric and not to a gravitational effect by which gravity acted directly on the photons of the light radiation.

However, the relationship between curvature of light and gravity spread in a generic way, as if photons were affected by gravity. This belief, without scientific basis, resulted in the subsequent appearance of optical phenomena called gravitational lenses, it was considered that its cause was due to the deviation of light by gravity, but as that assumption did not reach to explain the observed optical phenomenon, it was added as a complementary cause to the enigmatic theory of dark matter.

Both beliefs are absolutely false as we will demonstrate below:

As the General Theory of Relativity considers gravity to be a geometric phenomenon due to the model of the curvature of space-time contemplated in that theory and as we have previously demonstrated in this research that both things are wrong; accordingly by extension, we can confirm that the conclusion of Eddington's experiment is also wrong. Also obviously for the same reason we ruled out that the optical phenomenon called gravitational lenses has as cause to gravity.

But then if we dismiss gravity as causing light deviation in Eddington's experiment. What is the cause of this light deviation? There are two alternatives:

1) Light deviation is due to unknown agent
2) Light is observed with a curved trajectory in the universe

As there are no more elements involved in the experiment we consider the second option as true and we will demonstrate it:

10.11.1 Theory: "Light is observed by describing a curved trajectory in the universe"

Let's examine Figure 41

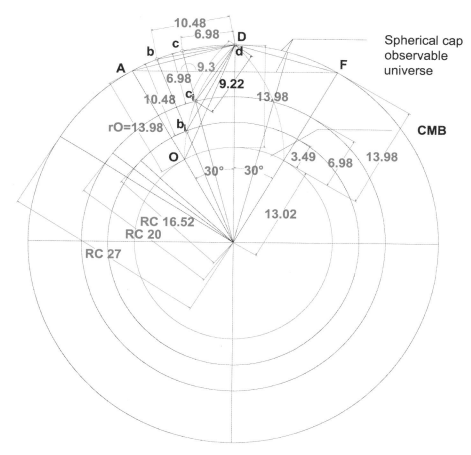

Figure 41 Curved light path   scale 1:10^9 light years

In Figure 41 we look at the vision bulb that we explained in chapter 3.1 of this research. As we have defined that matter is observed over time according to the shape of the vision bulb, necessarily the trajectory of the luminous signal will be following the profile of the bulb. This brings a conflict with the inviolability of the speed of light, because the shorter distance between two points is a straight line has no apparent explanation that the image of the ci galaxy moves to D following the trajectory of the c_i-D curve measuring 9.3×10^9 light years rather than moving by means of the c_i-D line measuring 9.22×10^9 light years. But this conflict has an explanation based on the relativity between space and time. The luminous signal that bears the c_i image, does not reach D directly or following the curve of 9.3×10^9 light years, nor following the 9.22×10^9 light years straight, The light arrives another way, as we will explain below:

When we look at the universe directly through a telescope we see images of the past of celestial bodies. We can't see the stars as they currently look. The farther away the oldest star is, the image we perceive of it. For example, if we look at from point D, located at its current site of the universe of $13,978 \times 10^9$ years, to Galaxy c, the image we observe of galaxy c is c_i, located in the universe of 6.98×10^9 years. This happens because the light that bears the image of c, takes 6.98×10^9 years to arrive from galaxy c to D. Similarly when we look at galaxy b located in the current universe, the image we observe is bi located in the 3.49×10^9-year-old universe because the light in the b image takes 10.48×10^9 years to reach D, which is the same time lapse it takes for the bi image to reach Galaxy b located in the current universe of $13,978 \times 10^9$ years-old. This means that the trajectory of light is always straight between galaxies located in the current universe, but they carry the image of the past when they were located in an older, smaller universe and therefore a radius of curvature less. The difference between the radius of curvature of the image site and our location produces a curvature in the path of the image. That curved deformation of the path image increases over time because the older the image, the greater the difference in the radius of curvature of the ancient universe of location, the image and the radius of curvature of the current universe.

In conclusion light has a straight trajectory in our current universe but is observed with a curved trajectory when we observe the celestial bodies of the universe located in the past and therefore in all astronomical observations the light will perceive curve increasing the curvature with the distance and hence the age of the observed image.

10.11.2 The explanation of the curvature of light observed in Arthur Eddington's experiment in 1919

In Figure 42 we will simulate Eddington's experiment:

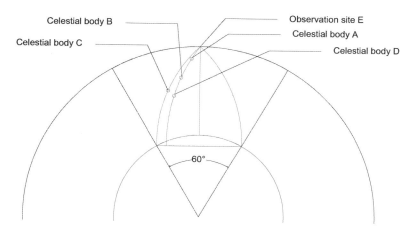

Figure 42 Light path of various celestial bodies

In this simulation the celestial body A represents the Sun and the celestial body C one of the stars that presented a curved trajectory during the observation of the eclipse. Due to the proximity of the Sun and the large magnitude of the radius of curvature of the light path 1,473 r0 the image of the Sun seems to reach the observer "E" straight, while that of the star much farther is perceived curve.

The result as is known was attributed to the deviation of the trajectory of light from the star by one of the walls of the curvature of the time space originated by the Sun. This experiment conducted by Arthur Eddington in 1919 had a great media impact and catapulted Einstein to fame and is considered the greatest astronomical test of the General Theory of Relativity.

However, we have found that this happened without the intervention of gravity and any other external factors. The above statement can be corroborated easily because we know that the sun does not create a concave curvature on which the planets of the solar system rotate. On the contrary, the ecliptic planes that form the Sun with its 8 planets are almost parallel, so there is no curvature in the solar system that can deflect light from a distant star.

10.11.3 The cause of the phenomenon called Gravitational Lenses

As we have previously ruled out that gravitational lenses are due to gravity and Dark Matter. You need to find your cause and that's what we'll do next: In Figure 42 we observed that celestial bodies A, B, and D are located on the same curve in the vision bulb. This results in the luminous signals of these three bodies being observed overlapping at the observation site. This overlay of luminous images produces optical phenomena such as those observed in gravitational lenses. These phenomena will be very varied depending on the dimensions and position that the celestial bodies have on the same curve or a very close curve. Multiple cases can be tested in an optical laboratory and try to reproduce some of the optical phenomena presented by Gravitational Lenses.

In conclusion the phenomenon called Gravitational Lenses is an optical phenomenon due to the curved trajectory of light being observed in space-time.

10.12 There is no anomaly in the mercury orbit

The idea that Mercury's orbit had an anomaly arose from the difference between the observed reality and the calculations on the perihelion precession of Mercury's orbit. The phenomenon of perihelion precession consists of a permanent rotation of the ecliptic plane of the planetary orbit, whose axis of rotation is the position of the planet in perihelion, this phenomenon was attributed to the disruptive effect of the other planets, Calculations based on Newton's theory gave results coincident with the observed reality, for all planets of the Solar System, with the exception of Mercury, in which case the calculations gave a value of 531"/century and astronomical observations indicated 574"/century, which yielded a remarkable discrepancy of 43"/century. That difference between the observed and calculated value could not be clarified for years, which created the myth of the supposed anomaly of Mercury's orbit.

In 1915 when World War I broke out, a German astronomer decided to enlist to participate in the war. But before he left at the helm, with the rank of lieutenant, Karl Schwarzschild reviews Einstein's theory of General Relativity and sends Einstein a simplified solution of his equations, hoping that Einstein would take into account his proposal. By that time Albert Einstein was already a renowned scientist, future Nobel Prize win-ner for his theories about light photons, however he was in an immense sense of frustration, because he was 8 years old trying to come up with a solution to his complicated equations of the Theory of Relativity Gen-eral.

Einstein welcomed Schwarzschild's proposal, as it was the first solution to his complicated equations, and I immediately used it in an intense two-week work to create a solution to the problem of Mercury Perihelion. The situation was ideal because if through his new General Theory of Rel-ativity he solved the mystery of Mercury's orbit, his theory would re-ceive worldwide recognition by solving a problem impossible to solve by Newton's theory. And he succeeded, according to Einstein Mercury because of its proximity to the Sun, suffered an effect in its orbit as a result of the photon's kinetic energy having to compensate for the gravitational energy exerted by the Sun in order to move away from it. Using the solu-tion of his equations provided by Schwarzschild, Einstein obtained an al-most exact 42,98" result from the missing 43" seconds that Newton's theo-ry could not calculate, because that theory did not consider the effect of light devi-ation. Accuracy is impressive, because it is a calculation involving large magnitudes in the data, such as the mass of the Sun, the mercury orbit and also the magnitude to be calculated is tiny of just 43 arc seconds in a century. However, despite the great margin of error of the data to which the margin of error of the observations must be added, the accuracy of Einstein's calculation was such that the 0.02" per centu-ry missing in its calculation, I justify it with the flattening of the Sun.

Einstein's success was total, as since the presentation of his solution to Mercury's problem, he received the personally great world renown and his general theory of Relativity was credited by the scientific community as the best theory to explain gravity, replacing Newton's old theory, which had been imposed for more than 200 years but was unable to analyze the effect of light deviation.

But then something strange happened, it was foreseeable that Newton's theory would be forgotten or relegated to the dumpster of history, due to the emergence of a new and superior theory, but it was not so, ???.

All calculations continued to be done using Newton's theory, with the ex-cuse that they are simpler than the complicated procedures required by the General Theory of Relativity. Einstein's theory won a single, but de-cisive, battle of Mercury. It was the only one that because of its com-plexity required to remove the heavy arsenal of the Theory of General Relativity and since there is only one Mercury in the solar system, there were no more battles.

In addition there were no more battles, because four years later in 1919 Arthur Eddington observed in an eclipse the deviation of light from distant stars which according to Einstein, confirmed his theory of the curvature of space-time by gravity, in this case the gravity of the Sun. That event received immense media coverage. New York Times gave Einstein the nickname "The Curved Man". There was no more doubt, Einstein's theory was definitely superior to Newton's and no one would ever dare challenge it, without the risk of being considered a heretic by the world of sci-ence.
In our research on a new alternative theory to the Standard Cosmological Model, we had an obligation to investigate all existing theories and especially Einstein's theories for their relevant role in astrophysics and because that theory was conceived when the expansion of the universe was not known. In the previous chapter we showed that Eddington's experiment what he actually revealed is that light is perceived curved by the curvature of the universe which is spherical and that light could not be deflected by the curvature of the space-time conceived by Einstein since according to that theory stars like the Sun create a deep pot on the plane of space-time on the bottom of which the Sun is located. Therefore in Eddington's experiment the light could not be deviated by the curva-ture of space time, just as the light cannot be deflected from cars driv-ing down a street by the hollow potholes of the track.
In previous chapters we have shown that Einstein's General Theory of Relativity must be dismissed for its inconsistency with the expansion of the Universe and with Kepler's 1st law, and that Newton's Theory of Universal Gravitation must be revived, with the aggregates of the cause of gravity and the exact value of the gravitational constant G. discovered in this research. However, how can we dismiss a theory that can solve a gravity problem, which cannot be solved by Newton's theory?
The only possible answer to the previous question is to find an explanation by classical physics of Mercury's perihelion problem. And that's what we're going to try to do next:
Because Mercury is the only planet to present this anomaly, the first thing you logically need to do is find out what differences there are be-tween Mercury and the other planets, and study how that difference can influence the problem. The most obvious difference between Mercury and the rest of the planets of the solar system is its proximity to the Sun and it was precisely where Einstein directed his research, since his the-ory is based on the energy of the deviation of photons from sunlight by the gravity of

Mercury which alters his orbit. But we must find another difference from Mercury with the rest of the planets related to the prob-lem and that can be explained by Newton's theory and fortunately that difference exists and it is also ideal to solve the problem.

10.12.1 The angular acceleration envisaged in Kepler's 2nd Law solves the apparent anomaly of Mercury's orbit

Mercury has, by far, the most eccentricity planetary orbit of the entire solar system. For this reason, in accordance with Kepler's 2nd Law, there is a great differential between its minimum orbital velocity in the apohelion and its maximum orbital velocity in perihelion. Therefore, in each cycle of Mercury's orbit, a great acceleration occurs when the planet goes from its position in the apohelion to its position in perihelion. We will calculate the angular acceleration of Mercury to see if it corresponds to the 43"/century missing from 574"/century observed in the perihelion precession of its orbit.

Mercury's angular acceleration per orbital cycle is as follows:

$$\alpha = \frac{\omega_p - \omega_a}{cycle} \quad (E10-16)$$

Where $\omega_p = V_p/r_p$. The angular velocity in perihelion is equal to the tangential velocity in perihelion divided by the radius of the orbit in perihelion

$\omega_a = V_a/r_a$. The angular velocity in the apohelion is equal to the tangential velocity in the apohelion divided by the radius of the orbit in the apohelion.

We have the following data queried on Wikipedia:

V_p = 53,703 Km/sec; V_a= 43,591 Km/sec; r_p=4,6 10^7 Km; r_a= 6,98 x 10^7 Km

Reemplazando:

$$\alpha = \frac{53.703}{4.6 x 10^7} - \frac{43{,}591}{6{,}98 x 10^7} = 5{,}43 \times 10^{-7} \, rad/cycle$$

We convert rad/cycle a arcsec/cycle

5,43 x10^{-7}rad/cycle=5,43 x 10^{-7}x 180/π x 60 x 60= 0,112"/cycle

Considering that the orbital mercury period is 0,241 years:

$$0{,}112"/cycle \times 100/0{,}2408 = 46{,}53"/century$$

$$\alpha/century = 46{,}53"/century$$

As a result of angular acceleration, a 46.53"/century rotation of the axis of the mercury orbit occurs, the axis of which is the baricenter of the perihelion. This equals the moments of the pair of forces created by mercury acceleration. As a result there is a movement of the Sun known as **"wobble"**, the magnitude of which will depend on the eccentricity of the planetary orbit. Because the other planets have very low eccentricity orbits, angular acceleration is virtually negligible.

The final balance is as follows:

The final balance is shown in Table 9.

Table 9

Calculate Precession

Source of precession	Arcosec/century
Attributed to the disturbance of Other Planets	531,63
Angular acceleration, Kepler`2° Law	46,53
Total calculated	578,16
Total observed	574.1
Difference calculation/observed	4,06= 0,69%

With the 46 arc-seg added by angular acceleration the frame is completed, with the observed and calculated values being equivalent. This reveals that there is no anomaly in Mercury's orbit, simply that one of the variables involved in the calculation followed Newton's law was not considered.

This surprising result has the following consequences.

Einstein's General Theory of Relativity's explanation of the anomaly of Mercury's Perihelion is misplaced, as it is the solution to a problem that never existed.

Newton's theory of gravitation regains its full universal validity, as it had been relegated, for 105 years, by the General Theory of relativity to a secondary role, for its apparent inability to explain the phenomenon of the precession of Mercury's perihelion.

Paradoxically. Einstein's elaborate theory of the anomaly of Mercury's orbit, which served for the consecration of his General Theory of Relativity, now after the preceding demonstration, becomes a definitive and unthinkable argument for his dismissal in the world of science.

In the next chapter these conclusions will be widely ratified.

10.13 The Sun's hidden orbits.

Newton's theory states that gravitational pull is due to two equal and opposite forces. Which we were able to fully demonstrate in Chapter 10 of this investigation. This means that the force and consequently the energy that the Sun's gravitational field transfers to a planet, is equal to the force and energy that the planet's gravitational field transfers to the Sun. We can check and calculate the energy that the sun transfers to the planets during the path of an orbital cycle, as we know the mass of the planets and the length of their orbit around the Sun. But what about the energy that the planet's gravitational field transfers to the Sun?

The above is a great question; there should be an orbit of its own in which the Sun develops in its path a kinetic energy equal to the energy developed by the planet through its orbit. Therefore, in accordance with Newton's theory, there has to be an equivalent solar orbit for each planetary orbit of the solar system. Therefore another great question arises: Where are these orbits?

This is a great mystery of the science of astrophysics, which has never been revealed, even never investigated in the 400 years of the validity of the heliocentric theory, which was born scientifically with the laws of Kepler.

But those orbits exist; they are simply hidden by another name: "Precession of perihelion" a rotating movement of all planetary orbits, attributed to the disturbance of the other orbits.

But in the precession of perihelion the Sun does not move. How can the Sun travel an orbit without moving and where those orbits are located?

We will answer both questions:

Movement is perhaps the most relative physical activity there is. We've all felt the sense of vertigo that produces the parking of a train next to one previously parked. It is not possible to distinguish whether our train is moving or the other previously parked. It doesn't work to move along a path or stand still and move the road at the same speed. This latest example happens every million times, in the vast number of gyms that exist in the world, with the use of treadmill machines. This system used by treadmills is very similar to that used by the Sun to move in the orbits it possesses with each of the planets. The process is as follows:

When the planet is located in the perihelion, the shortest distance between the planet and the Sun, there is an elliptical gravitational equipotential coming from the planet that crosses the axis of the Sun. At that very moment the planet begins a new rotating cycle following the path of its orbit, and the Sun continues or begins the path of its orbit, spinning the gravitational equipotential whose axis of rotation is the site that the planet occupies when it is located in the perihelion. This results in a rotating motion of the entire ecliptic plane where the planetary orbit is located. It is therefore an orbital system with two orbits: a rotating planetary orbit with the translation of the planet and a rotating solar orbit without movement of the Sun.

The kinetic energy used by the planet to travel its rotating orbit must be equal to the energy used by the Sun in traversing its spinning orbit. But another question arises as we know the orbits of the planets by astronomical observations, but how can we know the shape of the Sun's elliptical orbit?. The answer is that we have to calculate it and precisely in this research we find the tools to make that orbital calculation 1. In the next chapter we will calculate the rotating solar orbits of the 8 planets of the Solar System.

10.13.1   The spinning elliptical Orbits of the Sun

In chapter 10.5 of this research we were able to demonstrate that the value of G, the gravitational constant, is only a constant in spherical bodies, but in spheroidal bodies that value is no longer constant, it is necessary to multiply the value of G, by an angular variable K(Θ), whose maximum value is K=1 when .Θ= 0 and the radius vector is coincident with the major axis of the spheroid, and its value is minimal when .Θ=90° and the vector radius is coincident with the minor axis of the spheroid and the radius is. For any point the K value will be according to equation $K=r^2/a^2$ where "r"=the vector radius module and "a" the value of the semi-major spheroid semi-axis. This ratio is met proportionally equal within the spheroid or in the external spheroidal gravitational fields, as well as in the elliptical gravitational fields resulting from the intersection of the spheroidal bodies by the ecliptic plane when they are part of a stable orbit. We also demonstrated in that chapter that the K(90) value of the elliptical gravitational field formed in the ecliptic plane will depend on the geometry of the spheroid and the angle of inclination it has with the ecliptic plane. Finally in that study we were able to show that the value of K(90) is related to the eccentricity of the orbit according to the following equation:

$$K(90) = 1 - \varepsilon \qquad (E10-13)$$

We also demonstrated in this study that the Sun creates at its intersection with the plane of the ecliptic an ellipse of equal eccentricity as the ellipse created by the planet in that ecliptic plane and therefore at the beginning of an orbit from perihelion both bodies will be intertwined by their respective elliptical gravitational fields according to the lowest G value i.e. K(90) and will keep the coordinate axes of the gravitational fields parallel, throughout the orbital path. For best Figure let's review the orbital simulation of two spheroidal bodies performed in chapter 10.5 as shown in Figure 36.

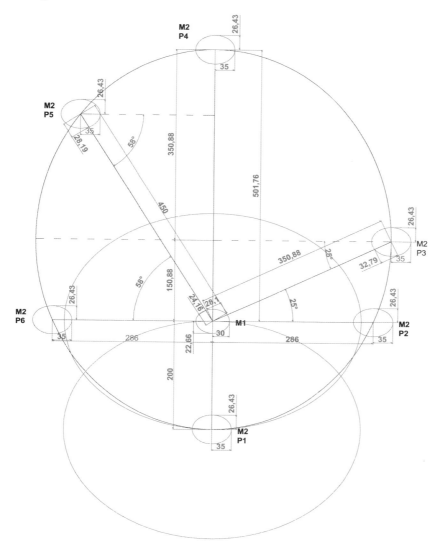

Figure 36. Orbit between two spheroidal bodies
M1 rotated body, M2 rotating body

In this Figure 36 we can observe the rotating elliptical orbit of the M2 rotary body with six path positions and the rotating orbit of the rotated body M1 consisting of the M2 gravitational field ellipse that passes through the center of the M1 body when both bodies are located in the perihelion. It is precisely at that moment that both orbital movements begin:

M2 Rotary Orbit: M2 travels along the orbital circuit in a time equivalent to its orbital period.

M1 Spinning Orbit: M1 begins the path of its orbit by rotating the M2 equipotential passing through the center of M1 as if it were a treadmill, whose axis is rotating is located in the center of the orbit at a distance of M1 equal to the perihelion of the rotating orbit. Consequently, the axis of the rotary orbit and the entire plane of the ecliptic occur. The cycle will be completed by rotating the 360° ecliptic plane.

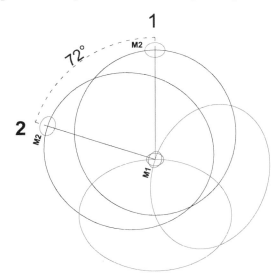

Figure 43 Spinning orbit of M1 at 20% of the period (72° of 360°)

We can calculate the solar spinning orbits of the eight planets in the solar system according to the following procedure:

To determine the dimensions of the orbit we only need the values of the semi-axis minor and the semi-major axis of the elliptical equipotential of the planet that crosses the center of the Sun in perihelion. The minor semi-axis we know and it is obviously the distance of the perihelion, it is only necessary to calculate the value of the semi-major field which we will do as follows:

We will calculate the value of the variable K(90) defined above using the equation K(90)=1-ε. The value of the eccentricity ε of each planetary orbit is well known. We'll take it in this Wikipedia case.

The value of the semi-major "a" of the orbit will be calculated using the following equation: $a = \sqrt{q^2/K(90)}$ (E10-15), where q is the distance of perihelion.

The mid-radius of the orbit will be calculated using the following equation: $Rms = \sqrt{(q^2 + a^2)/2}$ (E10-16).

To verify the veracity of this theory, we must calculate that the kinetic energy generated by the Sun traveling its orbit is equal to the kinetic energy generated by the planet through its orbit. Which we will do using the following equations:

Average speed of the planet: $Vmp = \sqrt{Msun \times G / Rmp}$ (E10-17)

Average speed of the Sun: $Vms = \sqrt{Mp \times G / Rms}$ (E10 – 18)

Kinetic energy of the planet: $Ep = 0.5\, Mp \times Vmp^2$ (E10-19)

Kinetic energy of the Sun: $Es = 0.5\, Msun \times Vms^2$ (E10-20)

Length of the planet's orbit: $Lop = 2\pi\, Rmp$ (E10 – 21)

Length of the Sun's orbit: $Los = 2\pi\, Rms$ (E10 – 22)

Energy used in the plane's orbit: $Eop = Ep \times Lop$ (E10-23)

Energy used in the Sun's orbit: $Eos = Es \times Los$ (E10-24)

To verify the veracity of this theory must be fulfilled, for the orbits of all the planets that Eop=Eos

Below are the tables (10, 11, 12 13) with the corresponding calculations made for the orbits of the Sun and the 8 planets of the Solar System.

Table 10

CALCULATION ORBITAL ENERGY OF THE SUN AND ITS PLANETS, COLUMNS A; B-F

A	B	C	D	E	F
Planet	Mass Kg	Half radius planet-Sun orbit (m)	Perihelion (m)	Apohelion (m)	Semi-minor axis planet-Sun orbit (m)
EARTH	5,97E+24	1,49598E+11	1,47098E+11	1,521E+11	1,49515E+11
MERCURY	3,30E+23	57894527007	46000567668	6,9817E+10	52768844679
MARS	6,42E+23	2,27937E+11	2,06669E+11	2,4921E+11	2,23932E+11
VENUS	4,87E+24	1,08209E+11	1,07476E+11	1,0894E+11	1,08199E+11
JUPITER	1,90E+27	7,77911E+11	7,40511E+11	8,1652E+11	7,74188E+11
SATURN	5,67E+26	1,42675E+12	1,35358E+12	1,5133E+12	1,41778E+12
URAN	8,69E+25	2,87097E+12	2,74894E+12	3,0044E+12	2,85958E+12
NEPTUNE	1,02E+26	4,45294E+12	4,50344E+12	4,5539E+12	4,45265E+12
PARAMETERS	SUN MASS	1,989E+30	G	6,67E-11	PHI
FORMULA/SOURCE		WIKIPEDIA	WIKIPEDIA	WIKIPEDIA	ROOT(C^2+(E-D)^2)

Table 11

CALCULATION ORBITAL ENERGY OF THE SUN AND ITS PLANETS, COLUMNS A;G-K

A	G	H	I	J	K
Planet	Semi-major axis planet-Sun orbit (m)	Eccentricity planet-Sun orbit	K(90)	Semi-major axis Sun Orbit (m)	Half radius Sun orbit (m)
EARTH	1,496E+11	0,0167	0,98329	1,48E+11	1,48E+11
MERCURY	5,7909E+10	0,2056	0,79437	5,16E+10	4,89E+10
MARS	2,2794E+11	0,0933	0,90669	2,17E+11	2,12E+11
VENUS	1,0821E+11	0,0068	0,99320	1,08E+11	1,08E+11
JUPITER	7,7852E+11	0,0484	0,95161	7,59E+11	7,50E+11
SATURN	1,4335E+12	0,0565	0,94352	1,39E+12	1,37E+12
URAN	2,8767E+12	0,0444	0,95559	2,81E+12	2,78E+12
NEPTUNE	4,5287E+12	0,0086	0,99141	4,52E+12	4,51E+12
PARAMETROS	3,14159265	SEG /DIA	86.400	°/RADIAN	57,29564553
FORMULA/SOURCE	(E+D)/2	WIKIPEDIA	1-H	ROOT(D^2/I)	ROOT((D^2+J^2)/2)

Table 12

CALCULATION ORBITAL ENERGY OF THE SUN AND ITS PLANETS, COLUMNS A; L-Q

A	L	O	P	Q
Planet	Mean tangential velocity planet (m/sec)	Mean tangential velocity Sun (m/sec)	Orbital period planet (dias)	Orbital period Sun (dias)
EARTH	29.779,47	51,93	365,32	206.848,67
MERCURY	47.869,79	21,23	87,95	167.495,83
MARS	24.125,33	14,21	687,08	1.084.285,96
VENUS	35.014,57	54,92	224,74	142.556,16
JUPITER	13.059,16	410,99	4.331,92	132.683,48
SATURN	9.642,86	165,90	10.759,91	602.169,04
URAN	6.797,76	45,65	30.713,50	4.430.188,60
NEPTUNE	5.458,29	38,91	59.327,52	8.435.563,26
FORMULA/SOURCE	ROOT(C13*E13/C)	ROOT(B*E13/K)	(2*G13*C)/(L*I13)	(2*G13*K)/(O*I13)

Table 13

CALCULATION ORBITAL ENERGY OF THE SUN AND ITS PLANETS, COLUMNS A; R-W

A	R	S	T	U	V	W
Planet	Kinetic energy Planet (Joule)	Kinetic energy Sun (Joule)	Planet orbit length (m)	Sun orbit length (m)	Path energy planet (Joule-m)	Path energy Sun (joule-m)
EARTH	2,65E+33	2,68E+33	9,40E+11	9,28E+11	2,49E+45	2,49E+45
MERCURY	3,78E+32	4,48E+32	3,64E+11	3,07E+11	1,38E+44	1,38E+44
MARS	1,87E+32	2,01E+32	1,43E+12	1,33E+12	2,68E+44	2,68E+44
VENUS	2,98E+33	3,00E+33	6,80E+11	6,76E+11	2,03E+45	2,03E+45
JUPITER	1,62E+35	1,68E+35	4,89E+12	4,71E+12	7,91E+47	7,91E+47
SATURN	2,64E+34	2,74E+34	8,96E+12	8,63E+12	2,36E+47	2,36E+47
URAN	2,01E+33	2,07E+33	1,80E+13	1,75E+13	3,62E+46	3,62E+46
NEPTUNE	1,53E+33	1,51E+33	2,80E+13	2,84E+13	4,27E+46	4,27E+46
FORMULA/SOURCE	0,5*B*L^2	0,5*C$13*O^2	2*G$13*C	2*G$13*K	R*T	S*U

The results indicated in **columns v** are identical to the corresponding ones indicated in the **w columns. so the theory of the Sun's spinning elliptical orbits is fully demonstrated.**

With the data from the tables (14,15,16,17,18,19,20,21,22) we draw the orbital systems of the 8 planets of the Solar System, figures (44,45,46,47,48,49,50,51,52):

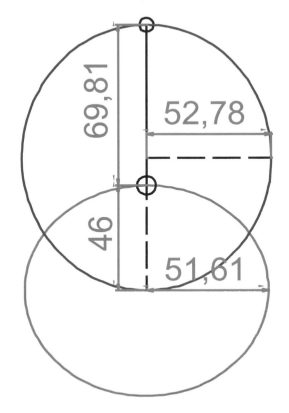

Figure 44 Mercury-Sun Orbital System

In blue orbit of the Mercury, in red orbit of the Sun, Scale: 1= million km

Table 14

Mercury-Sun Orbital System (values)

ORBITAL SYSTEM	TIPE	MEAT VELOCITY (m/sec)	PERIOD (Years)
MERCURY-SUN ORBIT	ROTATIONAL	47.869,79	0,24
SUN-MERCURY SUN	SPINNING	21,23	458,49

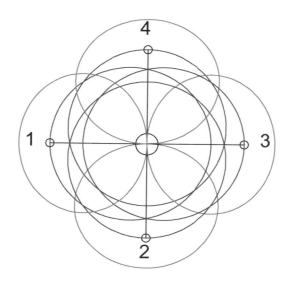

Figure 45: Four positions of Mercury's solar spinning orbit

P1= 25%, P2=50%; P3=75%, P4=100%

Table 15

Orbital values of 4 positions of the Sun-Mercury orbit

POSI-TION	% ADVANCE	TIME (YEARS)	DE-GREES	ADVANCE OBSERVED ON EARTH (DEGREES)	ADVANCE OBSERVED ON EARTH - (SECONDS)
1	25%	114,62	90	0,1589518	572,23
2	50%	229,24	180	0,3179036	1.144,45
3	75%	343,87	270	0,4768554	1.716,68
4	100%	458,49	360	0,6358071	2.288,91
ORBITAL PERIOD SUN-MERCURY YEARS)=			458,49		
ORBITAL PERIOD SUN EARTH (YEARS)=			566,21		
FOR THE OBSERVED ON THE EARTH WE MUST DIVIDE FOR HIS PERIOD= 566,21					

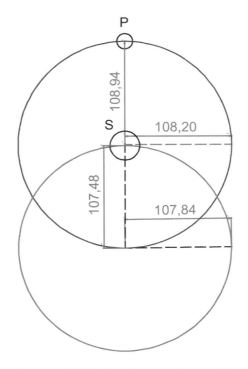

Figure 46 Orbital System Venus-Sol

In blue orbit of the Venus, in red orbit of the Sun, Scale: 1=million Km

Table 16

Venus-Sun Orbital System (values)

ORBITAL SYSTEM	TIPE	MEAT VELOCITY (m/sec)	PERIOD (Years)
VENUS-SUN ORBIT	ROTATIONAL	35.014,57	0,62
SUN ORBIT	SPINNING	54,92	363,20

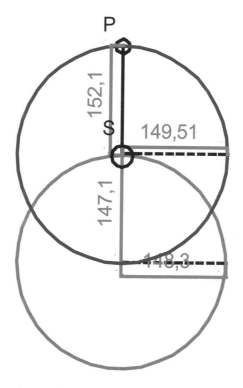

Figure 47 Orbital System Earth-Sun

In blue orbit of the Earth, in red orbit of the Sun, Scale: 1=million Km

Table 17

Earth-Sun Orbital System (values)

ORBITAL SYSTEM	TIPE	MEAT VELOCITY (m/sec)	PERIOD (Years)
EARTH-SUN ORBIT	ROTATIONAL	29.779,47	1,00
SUN ORBIT	SPINNING	51,93	566,21

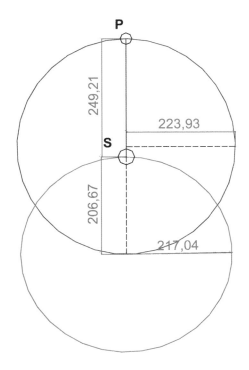

Figure 48 Orbital system Mars-Sun

In blue orbit of the Mars, in red orbit of the Sun Scale 1= million Km

Table 18

Mars-Sun Orbital System (values)

ORBITAL SYSTEM	TIPE	MEAT VELOCITY (m/sec)	PERIOD (Years)
MARS-SUN ORBIT	ROTATIONAL	24.125,33	1,88
SUN ORBIT	SPINNING	14,21	2.968,03

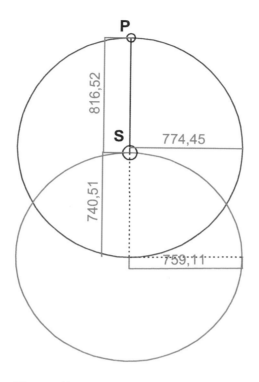

Figure 49 Orbital system Jupiter-Sun

In blue orbit of the Jupiter, in red orbit of the Sun, Scale: 1=million Km

Table 19

Jupiter-Sun Orbital System (values)

ORBITAL SYSTEM	TIPE	MEAT VELOCITY (m/sec)	PERIOD (Years)
JUPITER-SUN ORBIT	ROTATIONAL	13.059,16	11,86
SUN ORBIT	SPINNING	410,99	363,20

109

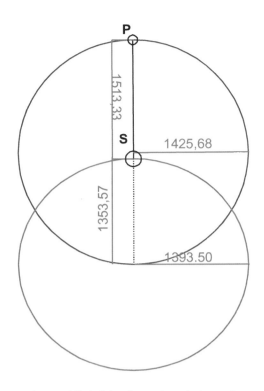

Figure 50 Orbital system Saturn-Sun

In blue orbit of the Saturn, in red orbit of the Sun : 1=million Km

Table 20

Saturn-Sun Orbital System (values)

ORBITAL SYSTEM	TIPE	MEAT VELOCITY (m/sec)	PERIOD (Years)
SATURN-SUN ORBIT	ROTATIONAL	9.642,86	29,45
SUN ORBIT	SPINNING	165,90	1.648,32

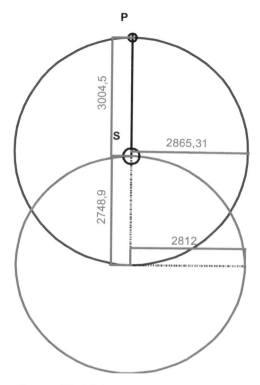

Figure 51 Orbital system Uranus-Sun

In blue orbit of the Uranus, in red orbit of the Sun, Scale: 1=million Km

Table 21

Uranus-Sun Orbital System (values)

ORBITAL SYSTEM	TIPE	MEAT VELOCITY (m/sec)	PERIOD (Years)
URANUS-SUN ORBIT	ROTATIONAL	6.797,76	84,07
SUN ORBIT	SPINNING	45,65	12.126,81

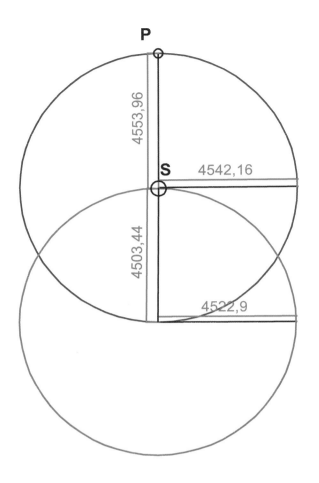

Figure 52 Orbital system Neptune-Sun

In blue orbit of the Neptune, in red orbit of the Sun: Scale:1=million Km

Table 22

Neptune-Sun Orbital System (values)

ORBITAL SYSTEM	TIPE	MEAT VELOCITY (m/sec)	PERIOD (Years)
NEPTUNE-SUM ORBIT	ROTATIONAL	5.458,29	162,40
SUN ORBIT	SPINNING	38,91	23.090,77

10.13.2 The cause of perihelion precession.

According to the Standard Cosmological Model, the precession of perihelion causes gravitational disturbances that are created between the rotating orbits of the planets. According to the Theory of the New Cosmological Model, the precession of perihelion corresponds to the spinning orbits of the Sun.

We will analyze the main differences between the two theories:

The disturbance generated by the gravitational pull of the rotating planets in the other orbits is dismissed as the cause of this phenomenon for the following reasons:

1) The orbits of the planets circulate in independent ecliptic planes.
2) Because of the large distance between planetary orbits, the gravitational attraction between the planets is negligible.
3) If there were any influence the same would be random, due to the large difference in orbital periods that the planetary orbits of the Solar System have. This is contradictory to the rotating movement of the ecliptic plane at constant speed presented by this phenomenon.
4) There is no explanation, for this case, of the origin of the energy required for this displacement of planetary orbits.

The cause of this movement according to the Theory of the New Cosmological Model attributed to the rotating orbits of the Sun, is clearly established in the law of universal gravitation, and mathematically proven by the calculations made in this chapter, which certify that the energy developed by each planet running through its orbit, is exactly equal to the energy used by the Sun to travel its rotating orbit around each planet.

10.13.3 Astronomical observations confirm the Theory of the Sun's spinning Orbits.

As an example we will analyze the case of the precession of Mercury's perihelion, a case famous for the supposed anomaly of its orbit. Case we studied extensively in the previous chapter:

Rotating Orbital Period of the Sun with Mercury: 458.49 years.
Period considered: 100 years.
The rotating advance will be the period considered divided by the orbital period multiplied by 360°:

$$Pr = \frac{100 \times 360}{458.49} = 78,52º/century \quad (E10-25)$$

The calculated value is very different from the observed value. However, because the observer is on Earth, which is spinning according to the speed of the Sun's rotating orbit with Earth, it is necessary to divide by the 566.21-year period of the Sun's rotating orbit with Earth. Therefore the real precession **Pr** is reduced being the observed precession **Po**:

$$Po = \frac{Pr}{566,21} = \frac{78,52º}{566,21} = 0,139º/century \quad (E10-26)$$

Converting degrees to arc seconds

$$Po = 0,139 \times 3.600"/º = 499.23"/century$$

To this value we must add the precession due to the angular acceleration of Mercury's orbit, calculated in the previous chapter, which is 46.53"/Century:

$$Pot = \frac{499,23" + 46,53"}{century} = 545,76"/century \quad (E10-27)$$

Compared to the amount calculated with the observed actual of 574"/Century, the result has an accuracy of 95.08%. This reveals a great accuracy, given the large magnitude of the observed period and the small value subject to the calculation, of a few seconds per century, to which we must add the error range of the observed values.

In this way it is clearly demonstrated, first, theoretically by Newton's law of Universal Gravitation, subsequently mathematically proven, and finally confirmed by astronomical observations, that the phenomenon known as perihelion precession corresponds to the rotating elliptical orbits that the Sun travels along together with each planetary orbit of the Solar system.

As a result of this unpublished discovery, we formulated our next theory

10.13.4 Theory. Modification of Kepler's laws

"The laws of Kepler shall be amended, to include solar orbits in such a way that it should be indicated that on each ecliptic plane, two similar orbits are created, one traveled by the planet and one traveled by the Sun".

This important discovery of solar orbits has far-reaching consequences for the theories that purport to explain to our universe.

1) It confirms all celestial bodies create a field of gravitational energy, proportional to their mass, which induces a force and energy proportional to the mass of celestial bodies that come into contact with that field. Therefore the same is true of the opposite, since

both bodies have a field of gravitational energy proportional to their mass.

2) This confirmation confirms Isaac Newton's theory of gravity in the law of universal gravitation and rules out Albert Einstein's theory of General Relativity, which considers that gravity is due to a deformation of space-time without the existence of a gravitational field.

3) The discard, previously demonstrated in the previous chapter, of the so-called mercury orbit anomaly is confirmed, as the phenomenon called perihelion precession has no relation to Mercury's orbit as it is due to the orbit developed by the Sun due to the energy induced by Mercury's gravitational field.

4) The only known difference between Mercury's orbit and the orbits of the other planets in the Solar system is the angular acceleration caused by the high eccentricity of its orbit. As a result of this angular acceleration there is a 46.53"/century rotation of the axis of the mercury. This matches the moments of the pair of forces created by Mercury's acceleration orbit, whose axis of rotation is the baricenter of the perihelion.

5) Consequently, the solution posed by the General Theory of Relativity, to an alleged anomaly of Mercury's orbit, is clearly wrong.

6) The theory of this research is confirmed, about the cause of gravity and the meaning of the gravitational constant G and its exact theoretical value.

7) The theory on the spheroidal gravitational fields of this research is confirmed.

11 THE BLACK HOLES.

Black holes are the most enigmatic elements of the known universe. They have been researched with great interest by the scientific community since the middle of the last century. However they remain enigmatic due to the amount of secrets they still hold about their existence. Let's look at the following definition of black hole by Wikipedia[11]

> A **black hole** is a region of space-time where gravity is so strong that nothing—no particles or even electromagnetic radiation such as light—can escape from it.[1] The theory of general relativity predicts that a sufficiently compact mass can deform space-time to form a black hole.[2][3] The boundary of no escape is called the event horizon. Although it has an enormous effect on the fate and circumstances of an object crossing it, according to general relativity it has no locally detectable features.[4] In many ways, a black hole acts like an ideal black body, as it reflects no light.[5][6] Moreover, quantum field theory in curved space-time predicts that event horizons emit Hawking radiation, with the same spectrum as a black body of a temperature inversely proportional to its mass. This temperature is on the order of billionths of a kelvin for black holes of stellar mass, making it essentially impossible to observe directly.
>
> Objects whose gravitational fields are too strong for light to escape were first considered in the 18th century by John Michell and Pierre-Simon Laplace.[7] The first modern solution of general relativity that would characterize a black hole was found by Karl Schwarzschild in 1916, and its interpretation as a region of space from which nothing can escape was first published by David Finkelstein in 1958. Black holes were long considered a mathematical curiosity; it was not until the 1960s that theoretical work showed they were a generic prediction of general relativity. The discovery of neutron stars by Jocelyn Bell Burnell in 1967 sparked interest in gravitationally collapsed compact objects as a possible astrophysical reality. The first black hole known as such was Cygnus X-1, identified by several researchers independently in 1971.[8][9]
>
> Black holes of stellar mass form when very massive stars collapse at the end of their life cycle. After a black hole has formed, it can continue to grow by absorbing mass from its surroundings. By absorbing

---

[11] Black hole - Wikipedia

other stars and merging with other black holes, supermassive black holes of millions of solar masses ($M_☉$) may form. There is consensus that supermassive black holes exist in the centers of most galaxies.............".

We can classify black holes according to their origin and according to their dimensions.

1) Stellar mass black hole[12] "They are formed when a star of more than 30-70 solar masses becomes a supernova and implode. They have more than three solar masses".
2) Intermediate black hole[13]:"(IMBH) is a class of black hole with a mass in the range of 100 to one million solar masses, significantly more than stellar black holes, but less than supermassive black holes".
3) Supermassive black hole[14]: with several million solar masses. They would be found in the heart of many galaxies. Chapter 6 of this research presents a theory about its formation.
4) Primordial black hole[15]: They were formed in the early universe by the direct collapse of a large portion of mass. Chapter 5 of this research presents a theory about its formation.

In the next chapters we will study each of these types of black holes:

12 STELLAR BLACK HOLE

When the implosion of a star of between 30 and 70 solar masses occurs, the immense explosion called Supernova destroys everything surrounding except the stellar remnant, in this case a black hole. What happens next to the position in space-time of the stellar remnant?. Before the implosion of the nucleus the mass that will constitute the black hole was within the nucleus at a distance from the bottom of the curvature of space-time equal to the radius of the star, as shown in Figure 53

Intuitively we could assume that the black hole would fall to the bottom of the curvature of space-time. But although the result of what happens is the same, it happens in another way, since it is the platform of space time that advances permanently at the speed of light, which reaches the site

---

[12] Stellar black hole - Wikipedia
[13] Intermediate-mass black hole - Wikipedia
[14] Supermassive black hole - Wikipedia
[15] Primordial black hole - Wikipedia

that occupied the core of the star, leaving the black hole at the bottom of the curvature of space time as shown in Figure 54

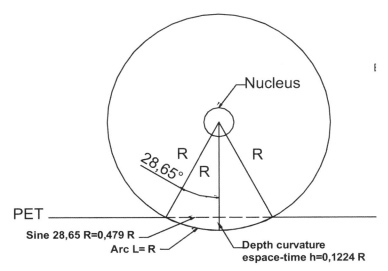

Figure 53 Star before the implosion of its core

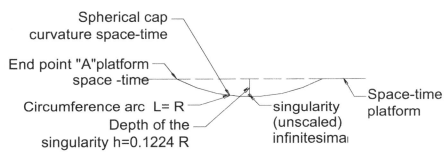

Figure 54 Location of the black hole's uniqueness after the implosion of the star's core

When making contact of the black hole with the background of the spherical cap of the curvature of space-time. The cap is transformed into an equivalent area cone. The dimensions of the cone will be:

Cap area: $A_{cap} = 2\pi R^2 (1-\cos 0,5\ radian)$ (E12-1)

Equivalent cone area: $A_{con} = \pi R\ sen\ 0,5\ radian \times G$; G: cone generatrix

Therfore:

$$A_{con} = A_{cap}$$
$$\pi R\ sen 0,5\ radian\ G = 2\pi R^2 (1-\cos 0.5\ radian)$$
$$G = 2R (1-\cos 0,5\ radian)/sen\ 0,5\ radian$$
$$G = 0,51\ R\ (E12-2)$$

Known the value of the Cone Generatrix and the Radius of the Base: $R\ sen\ 0,5$ radian, we get the depth of the cone to which will be located the singularity of the black hole.

$$h = \sqrt{(0,51)^2 - (sen\,0,5\;radian)^2} \times R^2$$
$$h = 0,1739\;R \quad (E12\text{-}3)$$

The new conical shape of the curvature of space time with the location of the singularity of the black hole, shown in the Figure 55

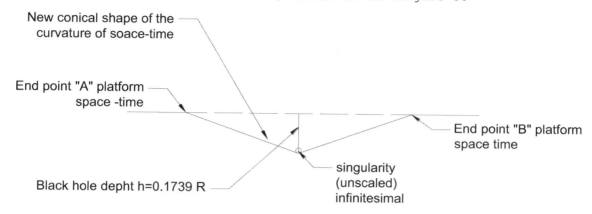

Figure 55 *Location of the singularity in the new conical shape of the curvature of space-time*

When the singularity, of infinitesimal dimensions, is located at the bottom of the cone, the mechanism of synchronized growth between the spherical mass of the collapsing star and the surface of space-time ceases. The loss of that synchronization brings the following problem: <u>The entire outer area of the cone will grow at the expense of an extra supply of space-time tissue.</u>

The extra space-time tissue flow needed to maintain the growth of the black hole's bottom cone area is obtained in two ways:

1) Radial flow of space-time: by growing the area of the cone due to the expansion of the universe according to Hubble's law.
2) Dynamic flow of space-time: through the principle of equivalence and the conservation of the angular momentum of the star that originated the black hole.

12.1 Radial flow of space-time

Consequently, due to the negative energy differential between the singularity and the plane of space-time, there is a drag of the tissue of space-time in all directions towards the mouth of the black hole's flujo radial, we quantify it as follows: Area of space-time required by the radial flow, as mentioned before, will be equal to the internal area of the cone, i.e.

$$A_c = \pi\,G\,r_e$$

Where G: generatrix cone=0.51 r_e; r_e: radius of entry to the black hole = 0,479R (E12-4), and R= redius of star

$$Ac = \pi \, 0{,}51\,R \times 0{,}479\,R$$

The gradient of that flow will be the Ac area, divided by the perimeter of the mouth:

$$\nabla Fr = \pi \, 0{,}51\,R \times 0{,}479\,R / 2\pi\, 0{,}479\,R$$

$$\nabla Fr = 0{,}255\,R \quad (E12\text{-}5)$$

La velocidad del flujo radial se obtiene dividiendo el gradiente entre el tiempo y como el tiempo es t=rO/C, tenemos:

$$Fr = 0{,}255\,R\,c/rO \quad (E\ 12\text{-}6)$$

Considering the great difference between the radius of a normal star and the radius of the observable universe, we can conclude that the radial flux is considerably slow in the case of stellar mass black holes.

Before calculating the dynamic flow of space-time we must define concepts and variables involved in its formulation:

12.2 Gravitational energy of a stellar mass black hole

The location of the black hole at the bottom of the cone of the curvature of space-time, has the consequence that it is delayed in the axis of time in relation to the level of the space-time platform and as the energy grows according to the equation $E = MC^2$ acquires a negative energy level, equivalent to the energy it would have if it had grown to the level of the space-time platform. Therefore the negative potential energy of the black hole can be calculated as follows:

$$E(bh) = -\Delta M C^2$$

$$E(bh) = -\Delta V \times \rho(bh) \times C^2 \quad (E12\text{-}7)$$

Where $\rho(bh)$ is the density of the black hole.

The growth volume is calculated using equations (E1-3):

$$x = (dx + c\,t)\ ; y = (dy + c\,t)\ ; z = (dz + c\,t); Z = (dZ + ct)$$

For the calculation of the volume on each of the coordinate axes, we will only consider the increase of the same:

$$x = (dx + c\,t - dx); y = (dy + c\,t - dy); z = (dz + c\,t - dz); Z = (dZ + ct - dZ)$$

For growth at the speed of light it is fulfilled that the growth of each axis is equal to the length traveled in hyperspace:

$$\Delta x = \Delta y = \Delta z = \Delta Z = ct = h = 0{,}1739\,R$$

Therefore ΔV will be a radio sphere $r = \frac{0{,}1739R}{2} = 0{,}08695\,R$

$$\Delta V = \frac{4\pi}{3}(0{,}08695\,R)^3$$

$$M = \Delta V \times \rho(bh) = \frac{4\pi}{3}(0{,}08695\,R)^3 \times D(bh) \quad (E12\text{-}8)$$

A virtual sphere of radius equal to 0.08695R and of density equal to that of the black hole.

Finally we obtain the negative or gravitational potential energy of the black hole:

$$Eg(bh) = -\frac{4\pi}{3}(0,08695\, R_e)^3 x\, \rho(bh)\, C^2$$

EQUATION (E12-9) OF THE GRAVITATIONAL ENERGY OF A STELLAR BLACK HOLE

(Where R is the radius of the star that originates the black hole)

The equation is valid for any star in the universe regardless of its size. Let's look at what Wikipedia says about the event horizon of a stellar black hole:

Wikipedia: Horizon of Events of a Stellar Black Hole [16]

Outline of the event horizon and ergosphere

"The event horizon is a spherically imagined surface surrounding a black hole, in which the escape velocity needed to move away from it coincides with the speed of light. Therefore, nothing within it, including photons, can escape due to the attraction of an extremely intense gravitational field.

The particles outside that fall within this region never come out again, because to do so they would need an escape velocity higher than that of light and, so far, the theory indicates that nothing can reach it.

Therefore, there is no way to observe the inside of the event horizon, nor to transmit information outward. This is why black holes have no

---

[16] https://es.wikipedia.org/wiki/Horizonte_de_sucesos

visible external characteristics of any kind, which allow to determine their interior structure or their contents, being impossible to establish in what state the matter is located since it exceeds the horizon of events until it collapses in the middle of the black hole. If we were to fall into a black hole, at the time of crossing the event horizon we would not notice any change, since it is not a material surface, but an imaginary border, away from the central area where the mass is concentrated. The peculiar feature of this border is that it represents the point of no return, from which there can be no other event but to fall inward, thus giving rise to the name of this surface. "

The quote above states that the event horizon is a virtual sphere that joins the black hole with the outside. Therefore, according to the principle of equivalence, the virtual sphere that equates to the gravitational potential energy of the black hole is the same sphere called the event horizon.

This means that the event horizon will interact with the outside with the potential properties of the black hole. Therefore we will calculate its virtual mass and its relationship to the outside, as if it were real.

$$M(eh) = \frac{4\pi}{3} (0,08695\, R)^3\, \rho(bh) \qquad (E12\text{-}10)$$

Where the radius of the sphere of the event horizon is:

$$r(eh) = 0,08695\, R \qquad (E12\text{-}11)$$

$$M_{hs} = 0,002753584\, R^3 x\, \rho(bh)$$

EQUATION (E12-12) FOR THE VIRTUAL MASS OF THE EVENT HORIZON OF A STELLAR BLACK HOLE

In the figure of the quotation appears an ellipse tangent to the event horizon called ergosphere. Let's find out what the Ergosphere is on Wikipedia

Wikipedia: Ergosphere[17]

"The ergosphere, also known as the ergosphere, is the outer region and close to the event horizon of a rotating black hole. In it, the black hole's gravity field rotates along with it dragging space-time.

---

[17] https://es.wikipedia.org/wiki/Ergosfera

It is a phenomenon theorized by New Zealand physicist Roy Kerr and emanates directly from Einstein's theories of general relativity. Kerr's black hole model is part of the first and simplest black hole model, Schwarzschild's model.

Its name was proposed in 1971 by Remo Ruffini and John Archibald Wheeler during the Les Houches conferences, and derives from the Greek word ergon, which means "work ». It received this name because it is theoretically possible to extract energy and mass from this region. The ergosphere has a flattened spheroidal shape that touches the event horizon at the poles of a rotating black hole and extends to a larger radius at the equator. The equatorial radius (maximum) of an ergosphere corresponds to Schwarzschild's radius of a non-rotating black hole; the polar radius (minimum) can be as small as half the radius of Schwarzschild in case the black hole is rotating to the maximum (at higher rates of rotation the black hole could not have formed).

Schwarzschild's model

Main article: Schwarzschild's Black Hole

The first fundamental model of a black hole was that of the German Karl Schwarzschild. Schwarzschild's black hole is basically a temporal singularity in null angular momenttime space-time and constitutes a simpler solution and the first of physical interest to be found to the equations of general relativity.

Kerr's model

Main article: Kerr's Black Hole

This model is a solution to the equations of general relativity for a rotating black hole. Such singularity, unlike Schwarzschild's, would have an annular form. The real black holes found in nature must be rotating as, by preserving angular momentum, they will rotate as the star or parent object did. It is known that the stars when they die lose much of the angular momentum, this being expelled along with the matter ejected by the supernova explosion in which the black hole forms. But, despite that loss of moment, some of it remains. Such a hole would produce, in a certain region called ergosphere, a "drag" area of space-time. The ergosphere is an ellipsoidal-shaped structure, coinciding its semi-minor axis with the axis of rotation of this. The ergosphere is therefore flattened in the direction of the axis of rotation in a similar way as the Earth does by the effect of its rotation.

The Ergosphere and Time Travel

Main Article: Time Travel: Time Travel

There is no rest inside the ergosphere. It is impossible for a body not to move, because the space itself revolves around the singularity so that the matter found in that region will rotate next to it. This fact according to the theory of relativity carries curious consequences. Observing a body traveling fast enough over the ergosphere could give a relative speed relative to us even higher than the speed of light c. In that case, such an object would simply disappear from our sight.

Penrose Process

Main article: Penrose Process

Because the ergosphere is outside the event horizon, objects from the gravitational pull of a black hole can escape in this region. An object can gain kinetic energy by entering the gravitational field of a rotating black hole and then escaping from it, taking with it some of the energy from the black hole. This process of energy absorption from a rotary black hole is called the Penrose Process and was developed in 1969 by mathematician Roger Penrose. [2] The theoretical maximum energy extraction that can be extracted is 29% of the total energy. When energy is absorbed, the black hole loses its turn and the ergosphere ceases to exist. This process is what could explain why black holes give off gamma rays. Computer models have shown that the Penrose process would be responsible for high-energy particle emissions being observed by quasars and other active nuclei of galaxies."

From the above quotation I would like to highlight, according to Kerr's Ergosphere model, the following aspects:

1) The real black holes found in nature must be rotating as, by preserving angular momentum, they will rotate as the star or parent object did.

2) It is known that the stars when they die lose much of the angular momentum, this being expelled along with the matter ejected by the supernova explosion in which the black hole forms. But, despite that loss of moment, some of it remains.

3) Such a hole would produce, in a certain region called ergosphere, a space-time "drag" zone

According to points 1) and 2) the black hole will retain part of the angular momentum of the star proportional to the mass of the black hole relative to the mass of the parent star, therefore we can estimate the turning speed of the black hole as follows:

The moment angles the star that will originate the black hole is equal to its moment of inertia multiplied by its angular velocity:

$$L_s = I_s \, \omega_s$$

The angular momentum preserved by the black hole shall be proportional to the mass preserved by the black hole

$$L_{bh} = L_s \, M_{bh} / M_s$$

So it's true:

$$I_s \, \omega_s \, M_{bh}/M_s = I_{an} \, \omega_{an}$$

Where: $I_s = 2/5 \, M_s \, r_s^2$ and $I_{bh} = 2/5 \, M_{bh} \, r_{bh}^2$

Substituting and simplifying:

$$r_s^2 \, \omega_s = r_{bh}^2 \, \omega_{bh}$$

Clearing:

$$\omega_{bh} = \omega_s \, r_s^2 / r_{bh}^2$$

The resulting angular velocity for the black hole will be of an almost infinite magnitude as it is proportional to the square of the radius of the star, which is very large, and inversely proportional to the square of the radius of the singularity, which is infinitesimal.

The angular velocity of the fictional sphere of the event horizon by equivalence will be:

$$\omega_{eh} = \omega_s \, r_s^2 / r_{eh}^2$$

Replacing with known values: $r_s = R$ and $r_{eh} = 0{,}08695 R$

$$\omega_{eh} = 1/(0{,}08695)^2 \, \omega_s$$

Result finally:

> $$\omega_{eh} = 132{,}27 \, \omega_s$$
>
> **EQUATION (E11-13) FOR THE ANGULAR VELOCITY OF THE EVENT HORIZON OF A STELLAR BLACK HOLE**

The result is that angular velocity increases for an observer trapped in the event horizon from **132,27** times the angular velocity of the primeval star to an almost infinite value to the bottom where the singularity rests. A tornado-like vortex.

12.3 Dimensions of the components of a stellar black hole

In Figures 56 and 57 we look at all the components and their dimensions of a black hole based on the magnitude of the primeval star.

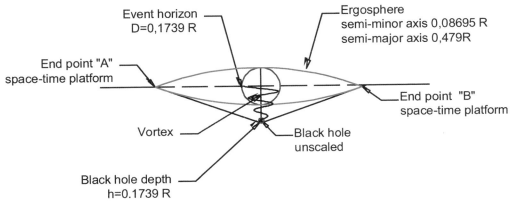

Figure 56 Side view and dimensions of the components of a stellar black hole

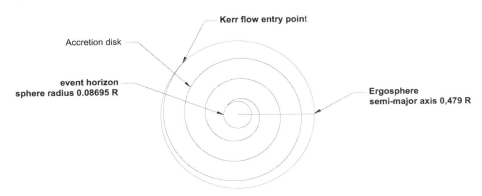

Figure 57 Top view and dimensions of the components of a stellar black hole

Point (3) of the above-mentioned Ergosphere model, says something transcendent: "Such a black hole would produce, in a certain region called the ergosphere, a drag zone of space-time".

This means that in some way, the vortex produced by the black hole will drag into space-time into the black hole. But there is a problem that needs to be solved in advance: The event horizon, which is the entry into the black hole vortex, is not connected to space-time, as it is at a distance of 0.479 R, from points "A" and "B", end of the space-time platform, as shown in Figure 55. How can you solve this problem?.

The solution is as follows, there are formations in the universe that do not settle in the space-time that are: Nebulae, interstellar gas, etc. which react to a movement like the vortex that produces the black hole and develop what is called a accretion disk[18]. Therefore, an accretion disk will be formed that will occupy transversely the spiral-shaped ergosphere perpendicular to the vortex formed by the black hole, surrounding the sphere of the event horizon, something resembling a hurricane containing in its eye or center the top of a twister. This creates a continuous flow of material to the bottom of the black hole. Once the accretion disk is formed with stellar dust, the disk comes into contact with the space-time platform and begins dragging that surface and the bodies placed on it.

Thus begins the permanent feeding of stellar black holes. It is a continuous flow of cosmic material at a constant rate, which we can calculate as follows:

$$F_k = \omega_{eh} \, r_{eh}$$
$$F_k = 132,27\omega_s \times 0,08695R_s = 11,5 \, \omega_s R_s$$

> **$F_k = 11,5 \, \omega_s \, R_s$**
>
> EQUATION (E12-14)) FOR THE SPEED OF THE TIME SPACE DRAG FLOW OR KERR FLOW OF A STAR BLACK HOLE

We have discovered an unknown phenomenon, until now, by physics, it is the flow of space-time into a stellar black hole. We have called it the Flow of Kerr, in recognition of New Zealand astrophysicist Roy Kerr, who forecasts his existence. Forecast that has been materialized and quantified according to the above equation, which indicates that the speed of the Kerr Flow is equal to the tangential velocity of the primigenial star multiplied by a dimensional constant of value 11,5, which we will also call Kerr's constant, K.

The discovery of this new physical variable is of enormous significance, as it establishes the relationship of black holes with the surrounding matter in a different way from that established by the law of universal gravitation.

The discovery of this new physical variable is of enormous importance, since it establishes the relationship of black holes with the surrounding matter in a different way than that established by the law of universal gravitation. This allows us to enunciate our new theory

---

[18] https://en.wikipedia.org/wiki/Accretion_disk

12.4 The flow of space-time into a stellar black hole, called kerr flow

All stellar black holes produce a flow of space-time into their interior which we will call, "Kerr flow", according to the equation $F_k = K \omega_s R$, being ω_s: the angular velocity of the primeval star, R: the radius of the primeval star and K, kerr's constant, with a dimensionless value of K=11.5. Matter near a black hole, regardless of the dimensions of its mass, will be subjected to a drag flow from the space-time where it is housed at a constant rate of 11.5 ω_s R in the direction of the black hole's ergosphere introduction site.

12.5 The acceleration of gravity of a stellar mass black hole

The gravitational energy of the stellar mass black hole equation (E12-3) creates a gravitational acceleration field with similarities and certain differences to that resulting from the law of universal gravitation:

1) The acceleration of gravity will be directly proportional to the virtual mass of the event horizon and inversely proportional to the distance from the center of the sphere of the event horizon (E12-4)
2) Stable elliptical orbits will not be created and therefore will not determine the trajectories of the masses affected by gravitational acceleration.
3) The trajectory of the mass particles into the black hole will only be determined by the helical trajectory of the Kerr flow.
4) The velocity of the mass particle on its path to the entrance of the black hole will be at the constant speed of the Kerr flow until gravity has a significant value, at which point its velocity is reduced to equal the speed due to gravitational attraction. From that point begins the zone of gravitational influence within which the velocity of the particle will accelerate inversely to the square of the distance to the center of the event horizon, but preserving the helical path of the Kerr Flow.

12.6 The phenomenon of pulverizing the attracted mass (tidal disruption event)

When the attracted object approaches the entry site of the accretion disk the effect of the gravitational pull force is so intense that it tears and pulverizes its mass. In this way the body enters and circulates in the accretion disk, in the form of cosmic dust of volume equivalent to that of its original mass.

12.7 Process of absorption of matter captured by the black hole

Matter in the form of cosmic dust circulates through the flat surface of the accretion disk in a helical trajectory increasing its speed, due to the gravitational pull force, until it reaches the sphere of the event horizon, where it comes into contact and is dragged by the vortex that forms between the event horizon and the black hole. Once inside the vortex it increases its speed of rotation to almost infinite values until it reaches its destination where it is absorbed and compacted by the black hole, increasing its density.

13 INTERMEDIATE-MASS BLACK HOLES

Intermediate mass holes (IMBH), with a mass between 100 and 1 million solar mass, were probably formed by multiple mergers of stellar mass black holes.

The Kerr flow produces a drag of matter located on space-time without exerting any force since there is no force that opposes it. However, a force could arise when two black holes approach and intertwine their respective Kerr flows, which would meet in opposition, creating two opposing forces.

13.1 "The drag force of space-time"

The drag force of each black hole would be given by the following equation:

$$\frac{dL}{dt} = r \times F$$

$L = I_{eh}\omega_{eh}$ y $r = r_{eh}$

Substituting and clearing

$$F = \frac{d\omega_{eh}}{dt} I_{eh}/r_{eh}$$

Since: $I_{eh} = \frac{2}{5} M_{eh} r_{eh}^2$; $M_{eh} = \frac{4\pi}{3} r_{eh}^3 \rho(bh)$; $r_{eh} = 0{,}08695\, R$ and the angular acceleration is: $\alpha_{eh} = \frac{d\omega_{eh}}{dt}$; In this case ω_{eh} it would cease to be a constant and vary according to the interaction of the forces. Therefore we have finally.

$$Fu_k = 95{,}77 \times 10^{-6} R^4\, \rho(bh)\, \alpha_{eh}$$

EQUATION (E13-1) FOR THE SPACE-TIME DRAG FORCE OR KERR FORCE

13.2 Process of joining two black holes

As two black holes approach, they can come into contact by inglating through their respective Kerr flows. The Kerr forces will then be produced according to equation (E13-1), which will be in opposition. They will approach attracted by the dominant force in helical rotating motions at constant speed until the corresponding spheres of influence of gravitational acceleration come into contact. From that moment the elliptical rotating motion of both holes will undergo a great acceleration to values close to the speed of light.

Finally the process concludes with the absorption of the smaller black hole creating a new black hole whose mass is equal to the summa the masses of the two black holes that join.

Intermediate-mass black holes (IMBH) were formed by repeated junctions such as the one described above.

14 SUPERMASSIVE BLACK HOLE

Supermassive Black Holes (SMBH) are located at the center of galaxies. In order to evaluate the behavior of supermassive black holes (SMBH), it is essential to know what galaxies are and how they work.

14.1 Basic information about galaxies

The Wikipedia article "Galaxy"[19]. It contains a good summary of the current theories of the Standard Cosmological Model, about their origins, classification. According to this article, galaxies are classified according to their shape according to the Hubble Sequence according to the diagram in the following image:

---

[19] File:Milky way profile.svg - Wikimedia Commons

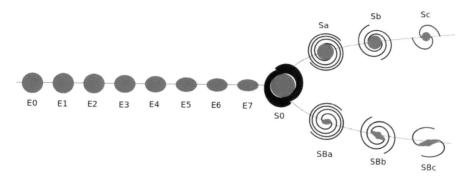

Figure 58 Wikipedia image with the classification of galaxies

E: Elliptical Galaxies; S: Spiral galaxies; SB: Barred spiral galaxies.
In relation to the component parts of a galaxy, they vary according to the type of galaxy, as an example below we can observe the components of the Milky Way according to the following Wikipedia image:

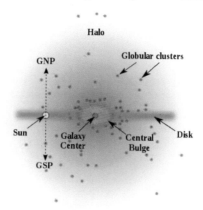

Figure 59 Wikipedia image with the components of the Milky Way

The articles cited only provide information about the shape and components of galaxies. The information about its origins is confusing and very scarce regarding the functioning and the relationship between its components since the Standard Cosmological Model assigns a determining role to the enigmatic and unknown theory of Dark Matter to the functioning of galaxies. It is mentioned that there is probably a supermassive black hole, inside its core, but it does not explain the influence of the SMBH on the functioning of the galaxy. And it could not be otherwise since in the Standard Cosmological Model there is no theory about the operation of black holes.

In Chapter 12 we elaborate the theory on the functioning of stellar black holes. In order to test whether this theory is valid for SMBH, we will try to explain the functioning of a galaxy by assuming that the SMBH of the galaxy behaves like a stellar black hole.

14.2 Analysis of the functioning of a galaxy based on the theory of stellar black holes

Figure 60 shows the surface of the spiral galaxy NGC 4258, also known as galaxy M106

Figure 60 showed the galactic bulge as a spheroidal bulb with a semi-major axis of approximately 10% of the diameter of the galactic.

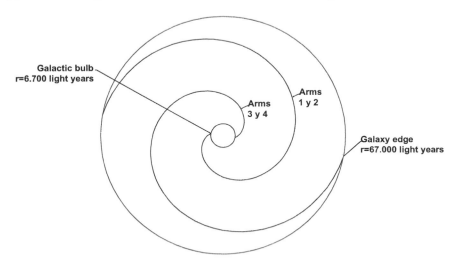

Figure 60 Galaxy NGC 4258

Dimensions of the galaxy, Galactic bulge radius equal to 10% radius of the galaxy, accretion disk r=1Ligth years

This spiral galaxy has been studied extensively, so we can check, based on the information that has been obtained from astronomical observations of it, if the theory exposed for stellar black holes is compatible with supermassive black holes.

Like all the 60 galaxies studied by Vera Rubin and Kent Ford, in the observations that served as the basis for the formulation of the dark matter theory, the stars of the galaxy NGC 4258 move at a constant high speed, from those located on the periphery to those located at a distance of 10% from the galactic radius, which coincides with the outer edge of the galactic bulge. From that point the speed of the stars is suddenly reduced until they reach the speed according to the Keplerian rotation.

As the SMBH accretion disk of the galaxy NGC 4258 is observed from the edge from Earth, it has been possible to make very precise measurements of its movement. The disk has clouds of molecular gas with the presence of rotating water massers. It is observed that the speed of displacement of the massers varies according to their position on the disk. The more internally located half a light-year from the event horizon move at a speed of 1,080 km/sec and take about 750 years to make a turn. On the other hand, the moreers located one light year from the center move at a speed of 770/Km/sec and take 2,100 years to make a turn.

We will check if the Keplerian rotation is fulfilled: The speed of the stable elliptical orbit of an object that rotates around a body of greater mass is calculated by the equation (E10-4):

$$v = \sqrt{GM/r} \quad (E10\text{-}4)$$

Therefore, two bodies rotating in independent elliptical orbits on the same body of greater mass must fulfill the following relationship between their velocities and their distances to the center of the rotated body:

$$\frac{v1^2}{v2^2} = \frac{\frac{GM}{r1}}{\frac{GM}{r2}} = \frac{r2}{r1} \quad (E14-1)$$

Replacing: v1=1080 Km/sec; r1=0.5 light years; v2=770 Km/sec; r2= 1 light years:

$$\frac{1080^2}{770^2} = 1.97 \cong \frac{1}{0.5} = 2$$

The Keplerian rotation is fulfilled since each máser behaves as if it were traveling an independent stable elliptical orbit following the law of universal gravitation. But this is not what happens, because both másers travel in the same disk of helical trajectory, in the direction of the black hole. Therefore this observation of the galaxy NGC 4258 is a strong demonstration of our theory which says: "Within the zone of influence of the gravitational sphere the Kerr flow defines the path of the bodies attracted by the black hole and the law of universal gravitation defines the speed of their translation"

14.3 Mass of the black hole of the galaxy NGC 4258

With the velocity data of the massers in the accretion disk we could calculate the equivalent mass of the black hole at the event horizon using equation (E10-4):

$$v = \sqrt{GM/r}$$

However the value of the gravitational constant G in this case is unknown. It is evident that it cannot correspond to the value of the universal constant of the celestial bodies.

However by way of reference we will calculate the value of the apparent mass of the event horizon considering the value of G =6,674x10⁻¹¹ N.m² Kg⁻²

$$M = v^2 r/G = (770 \times 10^3)^2 \times 1 \text{ light years}/G$$

$$M = 8{,}37 \times 10^{37} \text{Kg} \quad (E14-2)$$

Dividing by the mass of the Sun: M= 4.21×10^7 solar masses.

The equivalent mass of the event horizon of the black hole of the galaxy NGC 4258 would be 42.1 million solar masses.

14.4 Velocity of stars on the periphery of galaxies

The galactic bulge marks the boundary where the gravitational influence begins. In the rest of the galactic disk that makes up 90% of the galaxy's radius, the velocity of the stars does not correspond to a Keplerian rotation. The speed of the stars in that huge stretch of galaxies is constant and at a speed much higher than would correspond to a Keplerian rotation. The lack of a scientific explanation for this phenomenon gave rise to the theory of dark matter.

14.5 The Dark Theory of Dark Matter.

The theory of the Standard Cosmological Model, to explain the constant speed movement of stars in the outer stretch of galaxies, assumes as a cause the existence of an immense invisible mass. That extra invisible mass allows a formula used to measure the mass of galaxies, based on the velocity of stars, to match the velocity given by the formula with the observed constant velocity. The formula used was equation (E10-4):

$$v = \sqrt{GM/r} \quad (E10-4)$$

The formula (E10-4), as we explained above, is used to calculate the velocity of a massive body orbited by one of lower mass, when both form a stable orbit. However, in the case of the study carried out by Vera Rubin and Kent Ford in 1974, which gave rise to Dark Matter, its application was different. In that case the objective was to determine the mass of the galaxies, knowing the speed of the stars. At the time of Rubin and Ford's observations, the existence of supermassive black holes at the center of galaxies, nor their influence on the motion of their stars, was not known. Someone(?) assumed that stars in galaxies rotated in stable orbits according to the law of universal gravitation, in concentric circles but not around a common mass, but around the mass of the sum of the stars within their orbit. As an example it would be to suppose that the mass to calculate the speed of rotation of the Earth was not the Sun, but the sum of the mass of the Sun plus the masses of Mercury and Venus; or that the mass that determines the speed of rotation of Neptune would be the sum of the masses of the Sun and the other seven planets of the Solar System enclosed by their orbit. Which is an obviously erroneous application of the law of universal gravitation. Therefore, according to the proposed application of the erroneous formula, the larger the orbit, the greater the orbited mass. In this way the increase in the radius of the orbit in the denominator of the equation would be compensated by the increase in the mass enclosed by the orbit in the numerator of the equation. But a problem arose, in the outermost part of the galaxy the density of the star population decreased considerably, and therefore the mass enclosed in the orbit in the numerator of the formula could not compensate for the increase in the radius in the denominator, and therefore the stars of the periphery had to decrease their speed, which did not happen, since in the 60 galaxies observed the stars kept their speed constant until they reached 10% of the distance to the center of the galaxy from where it suddenly decreased drastically. As shown in the galactic rotation curve taken from Wikipedia.

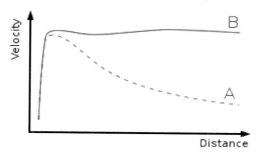

Figure 61 Galactic rotation curve (Wikipedia)

Curve "B" measured values. Curve "To expected values according to the formula used by the observers. Distance to the center of the galaxy.

Faced with these results contrary to what the equation used in the experiment predicted, the researchers, instead of concluding that this equation did not correspond to the observed reality and that, therefore, was not applicable for this case study, concluded the opposite. Everything indicates that the wrong equation was supposed to be right and the wrong was reality. Obviously absurd!

Which means that Dark Matter exists as the hidden mass necessary to modify the observed reality, so that the erroneous application of the Law of Universal Gravitation used in research can be fulfilled, in clear violation of the scientific method that establishes that theories must be formulated to explain the observed phenomena and not the opposite, that the observed must change to satisfy the fulfillment of a theory.

The most unusual thing is that a theory formulated about 60 years ago contrary to the scientific method to defend a clearly erroneous theory, instead of being discarded, has become a fundamental theory of the Standard Cosmological Model, which has expanded its use indiscriminately to explain any phenomenon, as is the case of intergalactic movement and the existence of gravitational lensing. It seems that the theory of Dark Matter acquired its own personality, as if it were really an invisible special matter, licensed to violate at will the scientific method in each of its applications.

When the scientific community finally reconsiders and the absurd theory of Dark Matter is definitively discarded, this unusual theory, which has cast so much shadow on scientific knowledge, will be sadly remembered as the greatest and longest error in the history of science.

The cause of the movement of stars on the outside of galaxies is an essential research task that must be undertaken as soon as possible by the scientific community. Task that presents 60 years of delay due to the shadow cast by the unusual theory of Dark Matter.

14.6 The Theory of Stellar Black Holes of the New Cosmological Model Explains the Motion of Stars on the Periphery and Inside galaxies

According to the theory of stellar black holes and therefore similarly in supermassive black holes bodies in the zone of influence of a black hole are attracted by the drag flow of space-time called Kerr flow and by gravitational attraction. Both phenomena act together but as gravity decreases with the square of the distance and the Kerr Flow is constant regardless of the distance, the farthest bodies are first attracted by it Kerr Flow at constant speed until reaching the zone of gravitational influence where a new stage begins in which the route follows the layout of the Kerr Flow, but speed is determined by gravitational attraction. Therefore the stars in the outer zone to the galactic core or bulge will move at the constant speed of the Kerr flow and in the inner zone at the speed according to the law of universal gravitation.

Therefore in the case of the galaxy NGC 4258 the velocity of the stars in the periphery should coincide with the value of the equation (E12-6)

$$F_k = 11,5\, \omega_e R_e \quad (E12\text{-}6)$$

In this case as the velocity of the stars on the periphery is known, we will verify the coherence of the factors of the formula. We do not have available the exact value of the speed of the stars on the periphery of the galaxy NGC 4258, therefore for the purposes of this theoretical exercise, we will assume the approximate value at the speed of the Sun in the Milky Way 240 Km/sec[20]. Therefore:

$$F_k = 11,5\ \omega_e R = 240 \text{Km/sec} \quad (E14-3)$$

As indicated in Chapter 12 the relationship between the radius of the accretion disk and the radius of the star that originated the black hole is as follows:

$$r_e = 0,479\ R \quad (E12-4)$$

As the radius of the accretion disk **Ra** is equal to **re**

$$R_a = 0,479\ R$$

$$R = R_a/0.479 \quad (14-3)$$

Replacing the radius value of the accretion disk of NGC 4258 Ra=1 light years

$$R = \frac{1\ ly}{0,479} = 2.087\ light\ year$$

$$R = 2,087\ lighr\ year$$

The radius of the star that originated the supermassive black hole of the galaxy averages 2,087 light year, which is a value consistent with the dimensions of the accretion disk of the galaxy of Ra = 1 light year. The angular velocity will be cleared from the equation (E12-6).

$$F_k = 11,5\ \omega_e R_e = 240 \text{Km/sec}$$

$$\omega_e = \frac{240 \times 10^3 m/sec}{11,5 \times 2,087\ light\ year}$$

$$\omega_e = \frac{240 \times 10^3\ m/sec}{11,5 \times 2,087 \times 9.46 \times 10^{12} m}$$

$$\omega_e = 1,056 \times 10^{-9}\ ciclo/sec$$

$$T = \frac{1}{\omega_e} = 946,9\ x10^6 sec = 33.25\ years \quad (E14-4)$$

Which means that the primeval star had a rotation period of 33.25 years, which is entirely consistent with the theory set forth in chapter 6 of this book, according to which the stars that formed the galaxies came from clones of primeval matter and therefore did not inherit the angular velocity of the accretion disks from which ordinary stars are formed. Verified the coherence of the data of the formula, we can confirm that the speed in the periphery of the galaxy NGC 4258 is due to the kerr flow of drag of space-time and of an approximate value of 240Km/sec. In chapter 14.2 we were able

---

[20] https://es.wikipedia.org/wiki/V%C3%ADa_L%C3%A1ctea

to verify that inside the galactic nucleus in the accretion disk of the SMBH the speed of the stars must correspond to a Keplerian rotation. Therefore it would be necessary to explain how the transition from one speed to the other is made; which we will do below: Previously we will define what the Galactic Bulb is:

14.7 Galactic bulb

The galactic bulb o bulge is a spheroidal body where the disk of the galactic nucleus is housed where in its center resides the SMBH, final destination of the stellar flow. The bulge is connected to the galactic arms and its function is to transfer the stellar flow from the galactic arms to the disk of the nucleus. Recent research[21] by an international scientific team led by the Center for Astrobiology confirms this theory. This study proves that the galactic bulge formed very early with the formation of galaxies, therefore it has an intrinsic origin in full agreement with the theory that we have exposed in this chapter and in total discrepancy with the theory of the origin of the galactic bulge established by the Standard Cosmological Model·, which attributes the origin of the galactic bulge to external causes[22] such as collisions and mergers between galaxies.

14.8 Process of transferring stellar flow from galactic arms to the disk of the galactic nucleus

The Kerr flow is a drag flow of space-time which generates a totally inertial movement at constant speed. That inertial motion remains constant on the outside of the galaxy in the 4 galactic arms where the gravitational pull produced by the equivalent mass of the event horizon is irrelevant because it decreases with the square of the distance. The four galactic arms are connected in pairs to the galactic bulge. Arms 1 and 2 at one point and arms 3 and 4 at the opposite end separated by a diameter of the bulb passing through the center of the galactic nucleus disk as shown in Figure 62. In both inputs there is angular momentum

$$L = r \, x \, mv$$

Where "r": Is the radius of spin=radius of the bulge and disk of the galactic nucleus; "v": entry velocity of the stellar flow of the galactic

---

[21] Identifican el origen de estructuras formadas en galaxias como la Vía Láctea - Ciencia - ABC Color
[22] Bulbo galáctico - Wikipedia, la enciclopedia libre

arms = 240 Km/sec and "m" the mass of the flow of stars. As a consequence, there is a rotational motion of the galaxy, whose axis of rotation is the center of the galactic nucleus where the SMBH is located.

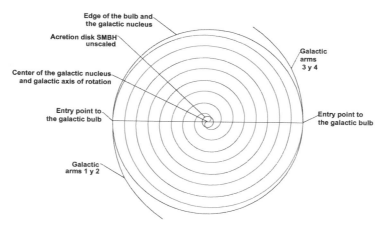

Figure 62 Helical trajectories of star flows, coming from the galactic arms, towards the SMBH

But at the edge of the nucleus the value of the gravitational pull force begins to be significant, which generates a conflict since stars traveling inertially at the constant speed of the Kerr flux of approximately 240 km / sec are at the same time attracted by the force of gravity at a speed of 9.407 km / sec, which is deduced using equation (E14-1)

$$v_2 = v_1\sqrt{r_2/r_1}$$

V_1= 770Km/sec, r_1= 1 light year; r_2= 6.700 light year; replacing values

$$v = 770 km/sec\sqrt{1/6.700}$$

$$V = 9,4 \text{ km/sec} \quad (E14-5)$$

It then happens that a body that travels inertially at a speed of 240 Km/sec, is added a force vector in the same direction that travels at a speed of 9.4 Km/sec. Consequently according to Newton's laws there will be a process of deceleration, by which the body reduces its inertial velocity of 240 km / sec to the speed of the force vector. The deceleration process is not instantaneous, with a braking space and time being necessary. The analytical calculation of the space required for braking is extremely complex due to the helical trajectory within the disk of the galactic nucleus, the uncertainty of the mass of the body and the resulting equalization velocity since the velocity according to the law of universal gravitation increases with the square in the direction of the trajectory towards the event horizon of the black hole. Therefore we will take the values obtained from the galactic rotation curve elaborated by the astronomical observations, which is shown in Figure 61 and we will particularize it for the case of the galaxy NGC 4258.

In Figure 63. We can observe the space velocity curve of the galaxy NGC 4258 where following the galactic rotation curve of Figure 61 the end point corresponds in the galaxy NGC 4258 at a distance of 3,350 light years, from the galactic center, whose speed using the galactic rotation curve corresponds to 13.3 km / sec and the same value is obtained using the equation (E14-1):

$$v_2 = v_1\sqrt{r_2/r_1}$$

v_1= 770Km/sec, r_1= 1 light year; r_2= 3.350 light years; Replacing values:

$$v_2 = 770 km/sec \sqrt{1/3.350}$$

$v_2 = v_b = 13,3$ km/sec (E14-5)

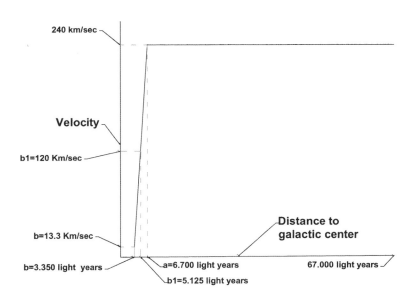

Figure 63 Galactic rotation curve. Galaxy NGC 4258

In Figure 64 we can observe inside the galactic bulge the braking process the flow of stars from the 4 arms of the galaxy NGC 4258. For a better explanation we will show separately the two trajectories, inside the nucleus disk.

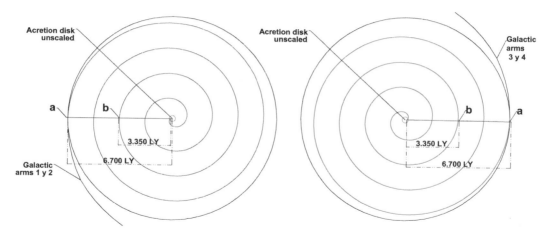

Figure 64 Galactic bulb galaxy NGC4258.

Braking path = helical length "a" to "b"; Vi=240 Km/sec, Vf=13.3 Km/sec

There are two helical trajectories one for arms 1 and 2 and another for arms 3 and 4 also helical but in reverse. However, both trajectories lead the stellar flow towards the accretion disk of the black hole located in the center of the galactic disk.

The process begins, in each case, at point "a", the entry site of the flow of the galactic arms, located at the edge of the bulge at a distance of 6,700 AL from the center of the galaxy and culminates at point "b" located 3,350 AL from the galactic center. The braking distance is equal to the helical length of the path between points "a" and "b". This process of changing the speed of the flow of stars, nebulae and other components of the galaxy has an enormous impact on its structure and functioning.

The change in speed of the stellar flow within the bulge of galaxies creates a traffic jam similar to that which occurs on vehicular traffic routes. As a result of the jam, an accumulation of stars, nebulae, stellar black hole planets and other components from the flow of the galactic arms is created. As in any jam, a tail or retention zone is created where the movement of the stars is very slow, therefore in that area the density of the stars and the illumination is maximum. We estimate that the retention zone in the galaxy NGC 4258 begins when the velocity drops to 120 km/sec, 50% of the speed of entry into the bulge, point b1, and ends at point b2 in the acceleration zone, where due to gravity, it increases again to 120km/sec. The number of stars in that retention zone remains constant, since for each new star that its velocity descends to 120km/sec and is incorporated into the retention zone, there will be a star in the acceleration segment that ascends to 120 km/sec and distances itself from the retention zone in the direction of the accretion disk of the black hole. To determine the magnitude of the retention zone we must calculate the coordinates with respect to the galactic center of points b1 and b2. For point b1 we will use the galactic rotation curve of Figure 63 and obtain the value corresponding to the speed of 120 km/sec; resulting in b1= 5,125 light years. To calculate point b2 we will use equation (E14-1):

$$r_2 = r_1 \frac{v_1^2}{v_2^2}$$

r_2=b2; v_2=120 Km/sec; v_1=770 km/sec; r1=1 light year. Replacing values:

$$b^2 = 1\ light\ year\ \frac{770^2}{120^2} = 41.17\ light\ years$$

$$b^2 = 41.17\ light\ years$$

In Figure 65 we observe of the helical trajectory of the galactic flux with the retention area with high star density and maximum illumination which corresponds to the path between point's b1 and b2. It occupies the central area in the form of an ellipse with the dimensions indicated: Major axis of the ellipse of 69% of the diameter of the nucleus and axis less than 63% of the diameter of the nucleus.

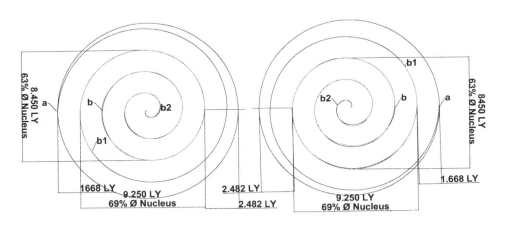

Figure 65 Star retention zone helical length from b1 to b2.

In Figure 66 we observe the illuminated elliptical zones resulting from the retentions of the flow of stars. On the left corresponds to the stellar flow of arms 1 and 2 and on the right to the stellar flow of arms 3 and 4.

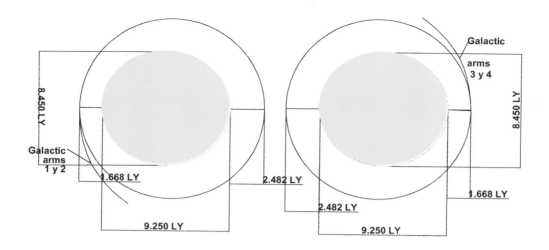

Figure 66 Ellipses of the lighting zone

In Figure 67 we observe the combined image of the illuminated part of both trajectories within the disk.

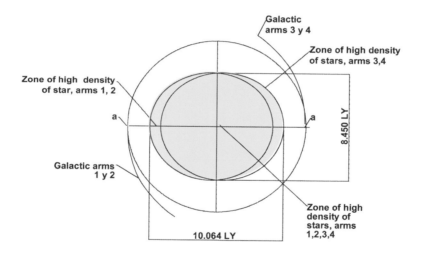

Figure 67 Illuminated area of the galactic nucleus of the spiral galaxy NGC 4258

14.9 The Bar of Barred Spiral Galaxies.

In barred spiral[23] galaxies the galactic arms are joined by an illuminated bar which rotates in an inclined plane.

---

[23] Galaxia espiral barrada - Wikipedia, la enciclopedia libre

Figure 68 Figura de Wikipedia: Galaxia espiral barrada NGC 1300

Assuming that the barred galaxy NGC 1,300 in Figure 68 had the same dimensions as the studio galaxy NGC 4258, the illumination zone of the galactic nucleus will correspond to the image in Figure 67 where the axis of rotation is aligned with the "z" axis orthogonal to the plane of the galactic nucleus disk. However, because the axis of rotation of the barred galaxy is in another plane, what is shown in Figure 69 happens.

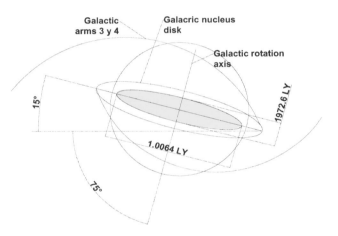

Figure 69 View of the disk of the galactic nucleus in an outdated plane 75° of the plane corresponding to a normal spiral galaxy. Maximum bar view

The disk of the nucleus is now in the new plane which is orthogonal to the new axis of rotation which is 75° out of time with respect to the previous one. As a consequence, one of the coordinates of the disk is reduced by its observable dimension by a factor equal to the sine of 75°. In Figure 69 the coordinate of the minor axis of the double ellipse of the illuminated nucleus is reduced and in Figure 70 the major axis is reduced by the same sine factor (75°). Due to the rotation of the disk plane the view of the galactic bar is permanently changing. The maximum size bar corresponds to figure 69 and the minimum size bar corresponds to figure 70.

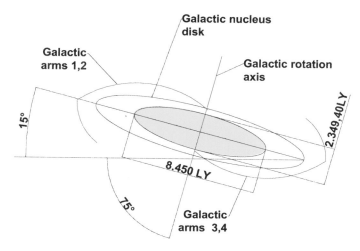

Figure 70 Minimum bar view

The observations of the Milky Way confirm the theory that we have just exposed about the bar of barred spiral galaxies, when they affirm the existence of a long bar and a short bar. It is not that there are two bars, one long and one short; What happens is that depending on the observer's site within the Milky Way he will observe the ellipse of the disk of a different size, between its maximum length of Figure 690 and its minimum length of Figure 70.

14.10 Illuminated area of the nucleus of elliptical galaxies.

Elliptical galaxies do not possess galactic arms, therefore they do not need or possess galactic bulge. Therefore the stars move from the periphery of the galaxy directly to the disk of the galactic nucleus. The process of slowing down the constant velocity of the periphery according to the Kerr flux as the galactic nucleus disk enters is similar to the process of spiral galaxies, but in a single helical trajectory within the nucleus, with a stellar flux that which is double that coming from a pair of galactic arms in spiral galaxies. Consequently in an elliptical galaxy of similar dimensions to NGC4258, the illuminated area of its nucleus will have the elliptical shape of one of the two images in Figure 66, but with twice the number of stars and twice their brightness, as shown in Figure 71:

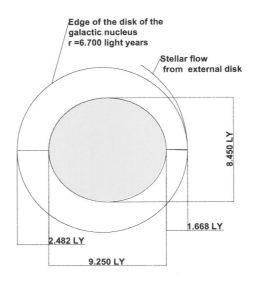

Figure 71 Image of the illuminated area of the nucleus of an elliptical galaxy of similar dimensions to the spiral galaxy NGC 4258

14.11 Final stretch of stellar flow

The final stretch of the stellar flux path takes place in the exit zone of the jam, which in the case of the galaxy NGC 4258 is 41 light years. Figure 72 shows the helical trajectory of the flow of the stars of arms 1 and 2 together with the helical trajectory but in reverse of the stellar flow of arms 3 and 4

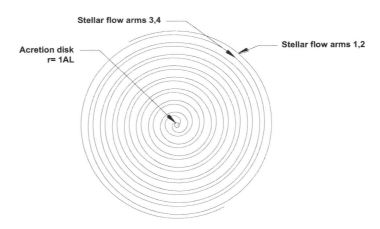

Figure 72 Final trajectory of stellar flow towards SMBH

Because the distances between the orbits are reduced as they approach the SMBH, multiple and complex gravitational phenomena can be generated by the effect of large stars and stellar black holes and IMBH, carried by both stellar flows. Finally the absorption process occurs.

14.12 Process of absorption of stellar flow by SMBH

In Figure 73 we can observe the final stretch of the trajectory of the stars of the galaxy.

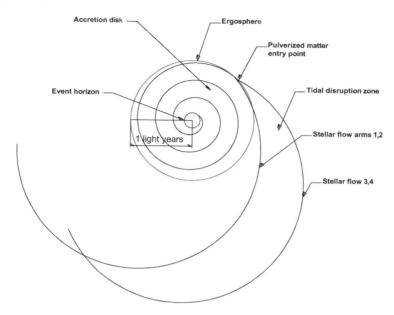

Figure 73 Accretion disk zone galaxy NGC 4258

In the vicinity of the accretion disk we indicate the estimated zone where the phenomenon of tidal interruption occurs, by which the stars are pulverized due to intense gravity, shortly before entering the accretion disk, to subsequently be absorbed by the black hole, similarly as Stellar Black Holes do, as described in Chapter 12.6. Astronomical observations confirm this[24]

14.13 Dimensions of the SMBH of the galaxy NGC 4258

The available information allows us to reveal the dimensions of the components of the supermassive black hole of the galaxy NGC 4258.

In Figure 56 we observe the components of a stellar black hole as a function of the radius of the original star. Knowing the radius of the accretion disk we obtained the radius of the star that originated the supermassive black hole by equation (E14-3)

$$Re = Ra/0.479 \quad (E14-3)$$
$$Re = 2.087 \text{ light years}$$

---

[24] https://www.lavanguardia.com/ciencia/20201013/484031785832/agujero-negro-devora-estrella.html

With the value obtained from the radius of the star we obtain the dimensions of the supermassive black hole of the galaxy NGC 4258 as shown in Figure 74

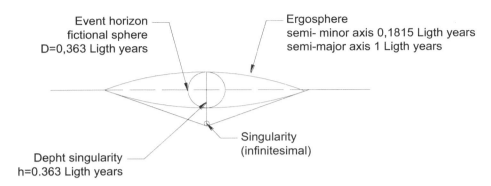

Figure 74 Dimensions of the components of the supermassive hole of the galaxy NGC42

15 PRIMORDIAL BLACK HOLES

Let's look at what Wikipedia says about primordial black holes:

> **Wikipedia: Primordial black hole[25]**
>
> *"**Primordial black holes** are a hypothetical type of black hole that formed soon after the Big Bang. In the early universe, high densities and heterogeneous conditions could have led sufficiently dense regions to undergo gravitational collapse, forming black holes. Yakov Borisovich Zel'dovich and Igor Dmitriyevich Novikov in 1966 first proposed the existence of such black holes.."*

Recent research by a group of scientists from Western University obtains relevant information on primordial black holes[26].

> *"Astrophysicists at Western University have found evidence for the direct formation of black holes that do not need to emerge from a star remnant. The production of black holes in the early universe, formed in this manner, may provide scientists with an explanation for the presence of extremely massive black holes at a very early stage in the history of our universe"*

According to the Theory of the New Cosmological Model, primordial black holes formed, as we explained in Chapter 5, at the beginning of the universe, when the spherical space-time caps of the Miniverses became tangent,

---

[25] Primordial black hole - Wikipedia
[26] Researchers decipher the history of supermassive black holes in the early universe - Media Relations (uwo.ca)

which produced an enormous contractionary force towards the center of each Miniverse. Due to compression a segment of the mass at the center of the caps became so dense that it exceeded the limits of compression at the quantum level similar to what happens with the formation of stellar black holes. As a result, a black hole was produced by the direct collapse of the mass in the center of each of the Miniverses, creating an inverted cone-shaped depression in the dome of the center of the spherical cap of the space-time of the miniverse, in the background of which the black hole was housed figures 75 and 76. We have assumed a conical surface similar in shape to the conical surface of stellar black holes in which the radius of entry to the trense is approximately three times the depth of the bottom where the black hole is housed. Since we do not know the dimensions of the pit, we will assume values as a fraction of the length of the observable universe. So the depth will be h= XrO and the radius of the inlet re=3XrO. X being a fraction of the observable universe. In the development of the study, based on observational evidence we will determine the value of X.

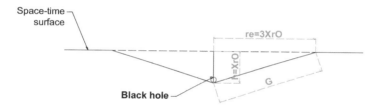

Figure 75 primordial black hole side view of the conical trenk

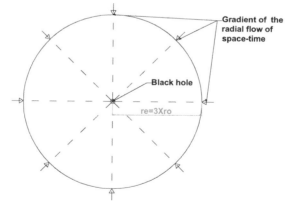

Figure 76 Primordial black hole top view of the conical trenk

As we have defined the components proportional to the radius of the observable universe, they will be maintained over time in the same proportion, since they will all grow in the same way by increasing the value of the observable universe radius **rO**. The mouth of the black hole located on the surface of space-time will grow according to Hubble's law as a fraction of the speed of light proportional to its diameter with respect to the radius of the observable universe, i.e.:

$$V\ growth\ (re) = \frac{3XrO x C}{rO} = 3XC$$

$$V growh\ (re) = 3XC \quad (E15-1)$$

But the surface of the cone of the pit is located below the platform of space time, therefore for its growth will be necessary an external flow of the mesh of space towards the interior of the pit. Consequently, due to the negative energy differential between the singularity and the plane of space-time, there is a drag of the tissue of space-time in all directions towards the mouth of the black hole. That radial flow, which we will call Primordial Flow, which we quantify as follows:

The area of space-time to be supplied by the primordial flow, as mentioned before, will be equal to the internal area of the cone, i.e.:

$$A_c = \pi\ G\ r_e \quad (E15-2)$$

Where the cone genaratrix $G = \sqrt{(XrO)^2 + (3XrO)^2} = 3,162\ XrO$; r_e: radius input to the black hole = $3XrO$

$$A_c = \pi\ 9.486(XrO)^2 \quad (E15-3)$$

The flow will converge in all directions; therefore we can calculate the gradient of the flow to the entrance of the black hole by dividing the required area by the perimeter of the inlet of the pit.

$$\nabla Fp = \pi 9.486(XrO)^2/2\pi 3XrO$$

$$\nabla Fp = 1.581\ XrO \quad (E15-4)$$

The gradient value for any other point x of the spherical cap of space-time will be obtained by dividing the area of the AC cone (E14-3) by the perimeter of the circumference whose radius Rx is equal to the distance between point x and the center of the miniverse.

$$\nabla Fp_x = \pi\ 9.486 \frac{(XrO)^2}{2\pi R_x}$$

$$\nabla Fp_x = 4.743 \frac{(XrO)^2}{R_x}$$

The velocity of that flow will be the value of the gradient divided by time. Time is equal to $t = rO/C$, therefore

$$Fp = 4.743\ X^2 rO^2/((R_x rO)/C)$$

$$Fp = \frac{4.743 x X^2 rO x C}{R_x} \quad (E15-5)$$

This is a problem similar to that of fluid mechanics, of draining a pond through a round drain, located in the center of the pond. The velocity of the fluid will be maximum at the entrance of the drain and will decrease inversely with the distance.

To complete the study it is necessary to determine the value of X, and substitute it in the equation (E15-5)

To prove this theory by practical physics it is necessary to find, in the observable universe, the site where the entrance mouth of a primordial black hole is located. the following conditions must be met at that site:

1) Clusters of galaxies, coming from all directions, should approach the site at high speed.
2) There must be a large empty space at the site, without stars or galaxies, which would correspond to the presence of the mouth of the primordial black hole.

Fortunately that site exists and is fully identified with The Dark Flow.

Two good descriptions of the Dark Flow can be found on Space.com[27] and Wikipedia[28].

From the quote from Space.com, we obtain the information that at the site observed at 6,000 million light years, multiple galaxy clusters move at the speed of 3.2 million km/h.

From the Wikipedia quote we highlight the following paragraph: "In June 2013 a cosmic map was published that includes background radiation data obtained by ESA's Planck telescope. This map shows a strong concentration in the southern half of the sky and a 'cold spot' that cannot be explained with the current laws of physics, but that would agree with the predictions of Mersini-Houghton and Holman. However, later studies have shown that it actually appears to be a huge area of the universe with hardly any galaxies."

Therefore the two conditions are clearly met to consider that the site, identified between the constellations of Sail and Centaur, houses the mouth of the primordial black hole.

From the quote from Space,com. we get the following values:

$F_d = 3,2 \times 10^6 \text{Km/h}$

We carry the speed indicated in Km/h to Km/sec:

---

[27] https://www.space.com/5878-mysterious-dark-flow-discovered-space.htm
[28] https://es.wikipedia.org/wiki/Flujo_oscuro#:~:text=En%20cosmolog%C3%ADa%20f%C3%AD-sica%2C%20el%20flujo,constelaciones%20de%20Vela%20y%20Centauro.

$$Fd = 3{,}2\ 10^6 \frac{Km}{h} x \frac{1h}{3600sec} = 888{,}88 \frac{Km}{sec}$$

$$Fd = 888{,}88 \frac{Km}{sec} x \frac{1C}{300.000\ Km/sec} = 2{,}96\ x10^{-3}C$$

Assuming that the observed site where the Dark Flow was measured was at a distance of 5% from the end of the path to the center of the miniverse we can calculate the corresponding X value using the equation (E15-5)

5% de 0.5rO= 0.025rO, therefore

$$Fp = \frac{4.743\ x\ X^2 rOxC}{R_x}$$

$$2{,}96\ x10^{-3}C = \frac{4.743\ X^2 rOxC}{0.025rO}$$

$$X = 3.945 x 10^{-3}$$

$$Fp_x = \frac{0{,}074 x 10^{-3} rO\ C}{r_x}$$

Therefore:

h=XrO=0.003945rO; re=3XrO=0.011835rO

$$h = 3{,}945 x\ 10^{-3} rO \quad (E15-6)$$

Depth of the primordial black hole trenk located in the center of each miniverse

$$r_e = 11{,}83\ x10^{-3} rO \quad (E\ 15\text{-}7)$$

Dimensions of the radius of the primordial black hole inlet located in the center of each Miniverse

$$Fp_x = \frac{0{,}074 x 10^{-3} rO\ C}{r_x} \quad (E15-8)$$

Equation for the speed of the drag flow of space-time, of a primordial black hole located at the center of each of the 56 Miniverses of the universe. r_x=distance to the center of the inlet of the primordial black hole; rO: radius of the observable universe; C speed of light

Replacing the value of re in the equation of the Primordial Flow at the entrance of the mouth of the black hole we have:

$$Fp = \frac{0{,}074 x 10^{-3} rO\ C}{11{,}83 x 10^{-3} rO} = 6{,}255 x 10^{-3}$$

$$F_p = 6,255 \times 10^{-3} C = 1.876,58 \, Km/sec \quad (E15-9)$$

Velocity of primordial flow at the entrance of the primordial black hole mouth located at the center of each miniverse

We have discovered a new phenomenon unknown, until now, by physics, it is the flow of space-time towards the interior of a primordial black hole.

This phenomenon which we have called Primordial Flow, governs the movement of clusters, clusters and superclusters of galaxies and leads them to the interior of the primordial black hole We therefore proceed to formulate our next theory:

15.1 The Drag flow of space-time into the interior of primordial black holes.

In the primordial black holes located at the center of each Miniverse, there is a flow of space-time that drags the formations of galaxies (Groups, Clusters and Superclusters) into the interior of those primordial black holes. The velocity of the now called Primordial Flow is calculated, according to the following equation:

$$F_{px} = \frac{0,074 \times 10^{-3} C \, r0}{r_x}$$

Where C is the speed of light, r0: the radius of the observable universe and r_x the distance from the measurement point to the center of the entrance to the primordial black hole.

15.2 The gravitational energy of a primordial black hole.

Just as we did with stellar black holes, we will proceed to calculate the negative potential energy of the primordial black hole.
The location of the black hole at the bottom of the curvature of space-time, has the consequence that it is delayed in the axis of time in relation to the level of the space-time platform and as the energy grows according to the equation E = MC² acquires a negative energy level, equivalent to the energy it would have if it had grown to the level of the space-time platform; i.e. the negative potential energy of the black hole can be calculated as follows:

$$E = -\Delta MC^2$$

$$E_p = -\Delta V x \rho(bh) \times C^2$$

Where ρ(bh) is the density of the black hole.

The growth volume is calculated using the equation (E1-3)

$$x = (dx + c\,t) \;;\; y = (dy + c\,t) \;;\; z = (dz + c\,t); Z = (dZ + ct)$$

Therefore:

$$c\,t = h = 0{,}003945\, r0$$

For the calculation of the value in each of the coordinate axes, we will only consider the increment of the same:

$$x = (dx + c\,t - dx) \;;\; y = (dy + c\,t - dy) \;;\; z = (dz + c\,t - dz); Z = (dZ + Ct - dZ)$$

For growth at the speed of light it is fulfilled that the growth of each axis is equal to the length of the displacement in hyperspace:

$$\Delta x = \Delta y = \Delta z = \Delta Z = ct = h = 0{,}003945\, r0$$

Therefore ΔV will be a radio sphere: r=h/2=0,001972 r0

$$\Delta V = \frac{4}{3} \pi\, (0{,}001972\, r0)^3$$

$$\Delta V = 82{,}62 \times 10^{-9}\, r0^3$$

The potential energy of the singularity will be the volume of potential growth multiplied by the speed of light squared and by the current density of the singularity, which is a variable in time

$$E_p = -\Delta V C^2 x \rho(t)$$

$$E_p = -\,82{,}62 \times 10^{-9}\, r0^3\, C^2 x \rho(t)$$

Therefore the gravitational energy differential between the singularity and the platform of space-time will be $Eg = E_p$

So we will finally have the definitive equation for the gravitational energy of a primordial black hole, which we will call the Big Bang Equation, for the reasons that we will explain later.

$$Eg = -82{,}62 \times 10^{-9}\, r0^3 x C^2 x \rho(t) \quad (E\ 15\text{-}10)$$

BIG BANG EQUATION

We have established the density of the singularity as a function of time ρ(t) since it increases over time for the following reasons:

- The Eg value increases over time according to the r0 value, which results in a permanent increase in pressure, with the consequent increase in the density and temperature of the singularity.
- The density increases with the clusters of galaxies that engulf the primordial black hole since its creation.

- Increasing the density of the singularity increases the value of Eg which in turn also increases again the density and temperature of the singularity. Establishing a virtuous circle of increasing the density and temperature of the singularity towards values that tend to infinity.

The value of the equation that we have called the Big Bang equation particularized to the current date will be:

How: $r0 = 13{,}978 \times 10^9$ light years

$$Eg = -82.62 \times 10^{-9} (13{,}978 \times 10^9 \text{light years})^3 \times C^2 \times \rho(t)$$

$$Eg = -1{,}15 \times 10^{21} (\text{light years})^3 \times C^2 \times \rho(t) \quad (E\ 15\text{-}11)$$

As shown in the equation, the negative energy differential that is established between the singularity and the space-time platform of a primordial black hole, is of a colossal magnitude, unimaginable by humans, which is also constantly growing at the speed of light. It is an energy that will undoubtedly have the capacity to produce, when the conditions come, an explosion similar to that of the Big Bang that created our universe.

This immense energy compresses the singularity of the primordial black hole, causing a permanent inflationary growth of its density and temperature tending both in time towards an infinite value, which allows us to affirm:
It is demonstrated, mathematically, that we are on an irreversible path to reproduce, in the 56 primordial black holes of our universe, the conditions of almost infinite temperature and density that the singularity of the Big Bang possessed, when the explosion that originated our universe occurred..

16 THE QUASARS

Quasars are the brightest astronomical phenomena in the universe and occur in supermassive black holes.

"WIKIPEDIA QUASAR [29]

---

[29] https://en.wikipedia.org/wiki/Quasar

Artist's rendering of the accretion disk in <u>ULAS J1120+0641</u>, a very distant quasar powered by a supermassive black hole with a mass two billion times that of the Sun

A quasar (/ˈkweɪzɑːr/; also known as a quasi-stellar object, abbreviated QSO) is an extremely luminous active galactic nucleus (AGN), in which a supermassive black hole with mass ranging from millions to billions of times the mass of the Sun is surrounded by a gaseous accretion disk. As gas in the disk falls towards the black hole, energy is released in the form of electromagnetic radiation, which can be observed across the electromagnetic spectrum. The power radiated by quasars is enormous; the most powerful quasars have luminosities thousands of times greater than a galaxy such as the Milky Way.[2][3] Usually, quasars are categorized as a subclass of the more general category of AGN. The redshifts of quasars are of cosmological origin.

The term quasar originated as a contraction of quasi-stellar [star-like] radio source – because quasars were first identified during the 1950s as sources of radio-wave emission of unknown physical origin – and when identified in photographic images at visible wavelengths, they resembled faint, star-like points of light. High-resolution images of quasars, particularly from the Hubble Space Telescope, have demonstrated that quasars occur in the centers of galaxies, and that some host galaxies are strongly interacting or merging galaxies.[5] As with other categories of AGN, the observed properties of a quasar depend on many factors, including the mass of the black hole, the rate of gas accretion, the orientation of the accretion disk relative to the observer, the presence or absence of a jet, and the degree of obscuration by gas and dust within the host galaxy".

16.1 Theory: The origin of quasars

One possible cause of quasar formation is the dragging radial flow of space-time, which is created in supermassive black holes. The radial flux is similar and is produced by the same causes as the one we have called in primordial black holes as Primordial Flow.

The equation for radial flow indicated in chapter 12 is as follows:

$$Fr = 0,255\, R\, c/rO \quad (E\, 12 - 6)$$

It is not possible to know the radius of the star that originated a stellar black hole. However we can measure the radius of the accretion disk that all black holes possess and since we know the relationship between the radius of the accretion disk and the radius of the primeval star, we can express the above equation as a function of the radius of the accretion disk. Consequently:

$$R_a = 0,479\, R$$
$$R = R_a/0,479$$

$$Fr = 0,532\, \frac{R_a}{rO}\, C \quad (E\, 16 - 1)$$

Obviously a very slow flow in the case of a star. However of a value also slow but more significant in the case of a supermassive black hole.

In Figure 77 we see that the radial flow enters through the edge of the ergosphere in a radial direction towards the sphere of the event horizon

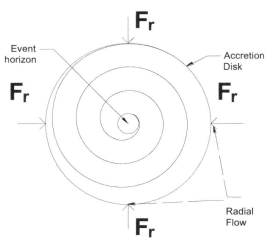

Figure 77 Radial E-T drag flow in a supermassive black hole

The radial direction and slow velocity of the radial flow conflicts with the helical direction and higher velocity of the Kerr flow in the accretion disk, thereby creating a propeller-shaped motion around the upper and lower part of the event horizon. The higher velocity of the kerr flow and its rotating coupling with the vortex leading to the black hole does not allow matter carried by the radial flow into the black hole to enter the black hole..

The propellers that form at the top and bottom of the event horizon acquire speeds close to that of light, bringing matter from the radial flow to a plasma state, which produces light radiation similar to that of stars..

The radial flow is slow, but the dragged matter remains indefinitely, unlike the matter carried by the Kerr flow which is completely absorbed by the black hole.

The material carried by the radial flow accumulates continuously during the billions of years of life of the galaxies, where they are housed. For this reason the propellers come to have dimensions greater than those of their galaxies.

16.2 Astronomical observations confirm the theory of the origin of quasars.

According to all astronomical observations quasars are more luminous and extensive in young galaxies, which are the oldest consequently. **The phenomenon is fading until it practically disappears in the nearest galaxies such as Andromeda and our own galaxy the Milky Way. This characteristic of increasing with distance and towards the older universe coincides exactly with the formula of radial flux (E12-6), which is as we have discussed before in this chapter the source of matter and stellar dust that feeds quasars.**

$$Fr = 0,255 \frac{R}{rO} C \qquad (E\ 12-6)$$

An analysis of the formula confirms this: The radius of the star divided by rO is the radius of the observable universe, a variable that increases linearly over time. Consequently the formula perfectly predicts the relationship of inverse proportionality with the time observed in the luminosity and extent of quasars.

17 THE INTERGALACTIC SYSTEM.

Galaxies are linked together to form galactic clusters. The way galactic clusters are created and moved, according to the Standard Cosmological Model Theory, is well described in the following Wikipedia citation:

Wikipedia:"Galactic grouping[30]

In the above quotation there is a full explanation of the formation and operation of the galactic groups. However, we disagree that galaxies relate to each other according to the Universal Gravitation Law.

Galaxies cannot relate to the Law of Universal gravitation among others for the following reasons:

As demonstrated in chapter 10 of this research, the gravitational relationship is established between two bodies. These two bodies by intertwining their gravitational fields and attracting each other, with two possibilities

1) If the required conditions are met, it will form a stable orbit.
2) If the conditions for a stable orbit are not met, the lower-mass body will crash into the higher-mass body.

In both cases it is a relationship only between two bodies. This means that the gravitational relationship between two galaxies would be determined by the individual relationship between the stars of each of them. This is impossible because of the following:

1) The enormous distance between galaxies does not allow the gravitational fields of stars to intertwine.
2) Even in the denied assumption that gravitational attraction can be established between several pairs of stars. The attraction of stars in the outer galaxy would have to overcome the gravitational pull that keeps stars within their own galaxy. This is impossible because the Law of Universal Gravitation states that the gravitational field is inversely proportional to the square of the distance.
3) A final option would be for a model to be created whereby the galaxy can act as an individual body with a mass equivalent to all stars and

---

[30] *https://es.wikipedia.org/wiki/Agrupaciones_galácticas*

with a gravitational field equal to the sum of the gravitational fields of all its stars. In that case the gravitational attraction between the equivalent models of both galaxies could be established. Although this can be done correctly in other cases such as the study of electrical circuits using Thevenin's theorem, it is impossible in the case of galaxies since the gravitational fields are individual of each body and do not mix.

For all these reasons, the gravitational attraction between galaxies is totally ruled out.

As established by the Theory of the New Cosmological Model in chapter 15 of this research the motion of galaxies is determined by the radial flow of space-time called Primordial Flow of primordial black holes and not by the Law of Universal Gravitation.

Consequently, to verify the veracity of this theory it will be necessary to demonstrate that primordial flux is capable of creating galactic clusters and determining the dynamics and evolution of them without the need for the Law of Universal Gravitation and the enigmatic theory of dark matter, and that is what we will try to do next.

17.1 Theory: Primordial Flux determines the motion of galaxies

According to the Theory of the New Cosmological Model, there is a primordial black hole at the center of each Miniverse. This primordial black hole creates the Primordial Drag Flow of space-time, which carries the Super-clusters of galaxies towards the mouth of the primordial black hole and is responsible for all intergalactic motion.

In chapter 15 we define the formula of the rate of drag of space-time by the primordial flow, which is as follows:

$$Fp_x = \frac{0,074 x 10^{-3} r0\ C}{r_x} \quad (E15-8)$$

The formula indicates that the drag speed is inversely proportional to the distance to the center of the mouth of the primordial black hole (PBH). It follows from the above that the speed will increase in smaller and smaller circles until it reaches the edge of the mouth of the PBH. As shown in Figure 78. Galaxies are carried by the dragging motion of space-time where they are installed in flow lines in a radial direction towards the center of the mouth of the primordial black hole (PBH) Therefore the galaxies will increase their speed and narrow the distances between them as they approach the mouth of the PBH.

The dimensions of the radius of the primordial black hole mouth are calculated by the equation (E15-7):

$$r_e = 11,83 \times 10^{-3} rO \quad (E\ 15\text{-}7)$$

Figure 78 shows the spherical cap of one of the 56 Miniverses in our universe. Whose radius is 0.5 rO.

On the periphery of the spherical cap of each Miniverse continues the permanent process, initiated at the beginning of the universe of the creation of space-time, implantation of clones of primitive matter and its evolution to constitute new galaxies.

Therefore in Figure 78, we can graphically see the entire vital process of matter in the universe from its creation to its final recycling, as it is absorbed by the Primordial Black Hole located at the center of each Miniverse.

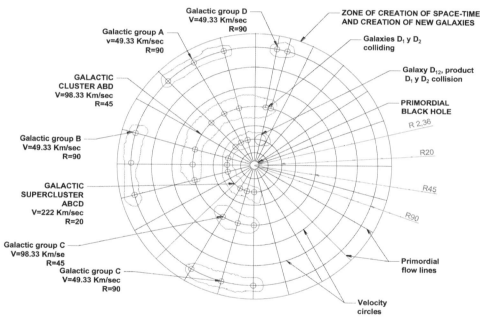

Figure 78 Primordial Flow in the spherical Cap of a Miniverse
Scale 200:1 Maximum radius 100=0.5 rO

17.2 Motion simulation of galactic groups

The 4 galactic groups A, B, C and D (Figure 78), depart from the periphery of the Miniverse at a distance from the mouth of the PBH of 0.45 rO, which corresponds to the circle of radius R 90 according to the scale of the drawing and a speed of 49.33 Km/sec calculated by the formula of the Primordial Flow Fp equation (E15-7), indicated at the beginning of this chapter.

Galaxies of all groups will move, always following the drag flow line of space-time where they are located. Because all flow lines are convergent at the center of the mouth of the primordial black hole, galaxies as they move forward approach each other, and galactic groups also approach each other. As a result, although it does not really exist, it seems that a force of attraction is created between galaxies of the same group and a force of attraction between groups of galaxies closer to each other. In this way in the path there is a continuous increase in speed and a continuous reduction in the distance between galaxies and between galactic groups

The four galactic groups upon reaching the R45 circumference experience the following:

1) In group D its two galaxies collide due to the proximity of their converging flow lines.
2) The galactic groups A, B and D have come so close that they become the ABD galactic cluster and increase its speed which reaches 98.66 km/sec.
3) The galactic group C, being further away, remains individual and also increases its speed to 98.66 Km/sec.

When the galactic groups reach the circumference R20, the following happens:

1) The distances between the galactic groups are reduced so much that the galactic group C, joins the ABD Cluster and form the ABCD Super-cluster, with greater mass and greater speed, which reaches 222 Km / sec.
2) In group D its two galaxies D1 and D2 have completed their fusion into a larger galaxy D_{12}

The clusters and Superclusters have been located for better visualization of the experiment at a distance greater than that which corresponds to the center of the mouth of the PBH, hence the low speed they present.

The end of the trajectory concludes at the circumference R 2.366, which is the mouth of the primordial black hole, where all the now much more compact groups coming from all directions are absorbed by the ANP at the maximum speed of 1,876.58 Km/sec. In this way, the astronomical observed in the phenomenon called Dark Flow is fully reproduced.

17.3 The primordial flux is not altered by the acceleration of gravity due to the gravitational energy of the primordial black hole.

Contrary to what happens with rotating black holes the velocity in the trajectory is not increased by the gravitational energy of the equivalent mass of the primordial black hole. The expansion of the universe renders ineffective the acceleration of gravity produced by gravitational energy, because the distance increases at a speed proportional to the distance to the center of the Miniverse, according to the equation (E8-3).

$$Vab = \frac{X}{rO} \, x \, C \quad (E8-3)$$

Where "a" in this case is the center of the miniverse "b" the site of the trajectory and "X" the distance between the two points. Only in the vicinity of the mouth of the black hole where the distance between "a" and "b" is less are the velocities equalized and the effect of tidal disruption occurs due to the excessive gravity of the primordial black hole.

17.4 System of recycling of the matter of the universe

The process we have described has a function of continuous recycling of the matter of the universe, since all the matter of the universe is inside the galaxies. Therefore the older galaxies are continuously absorbed by the Primordial Black Hole of each Miniverse. This produces a continuous deficit of the total mass of the Miniverse. As the mass of the Miniverses must increase continuously due to equation (E3-1), the replacement of the absorbed mass by a ring with clones of the primordial matter on the edge of each Miniverse occurs in a synchronized manner, the number of clones of which will be necessary to compensate for the recycled matter and increase the total mass of each Miniverse due to the increase in energy derived from the expansion of the universe. It is a continuous process, by which the elimination of the oldest galaxies, gives way to the formation of new galaxies arising from replicas of the primordial matter of the universe.

17.5 Evaluation of the experiment of simulation of the movement of galactic groups

In the previous Wikipedia quote on galactic clusters we highlighted the aspects that we believe to be the most important that derive from the astronomical observations made regarding the behavior of galactic clusters. We will evaluate whether our theory of the motion and operation of galaxies and their clusters meet those observations:

1) When visually observed, clusters appear as collections of galaxies self-sustaining gravitational attraction. However, their velocities are too great to remain gravitationally limited by their mutual forces of attraction. This observation demonstrates the implication of the presence of an invisible additional component.

2) The more mass the cluster has, the higher the escape velocity. Also, more mass implies greater gravitational forces, which leads to higher accelerations and higher speeds. Thus, in the most massive clusters the galaxies that compose them move faster with respect to one another than in the less massive ones.

3) Some tend to concentrate more matter by adding small clusters and other individual galaxies, which leads them to compact more and more.

4) Groups, clusters and some isolated galaxies can form larger structures: superclusters. These clusters would behave similarly to clusters, only in them the elementary particles that constitute them would no longer be individual galaxies, but entire galactic groups and clusters that move confined to their colossal gravitational field.

According to the results of the experiment, it is evident that the galactic motion model of the NCM broadly and accurately complies with the provisions of the 4 paragraphs of the quotation, but with many advantages over the theory of the Standard Cosmological Model (ECM).

1) The galactic motion model of the NCM does not use gravity, according to the Law of Universal Gravitation, which we have shown does not work in this case and does not use the enigmatic theory of Dark Matter.

2) In the case of the New Cosmological Model (NCM) galaxies have a clear destiny in their motion, which is to be absorbed by the primordial black hole through the recycling system described in chapter 17-4, something that is essential in any system. In the case of the ECM, the motion of galaxies has no known destination, one might ask: do they wander aimlessly defined forever?

3) The theory of gravitational attraction as a cause of the displacement and approach of galaxies, in addition to being impossible as we have shown above, in the denied assumption that it worked, is unfeasible, by the following reasoning that inevitably reaches an absurdity: According to this theory a galactic cluster owes its movement to the gravitational attraction of another larger one, that in turn, must

be attracted by another even larger one and so on, in a series that would lead to an infinite number of clusters of galaxies, and since there is no recycling, the chain has no end , therefore the largest of those galactic clusters would be of infinite size.

Well we have explained how the intergalactic system works in a Minverse, does the same thing happen in the rest of the universe?. The answer is obvious, of course it is, but separately and individually in the other 55 twin Miniverses. In Figures 79 and 80, we look at our Miniverse and its 6 tangent Miniverses:

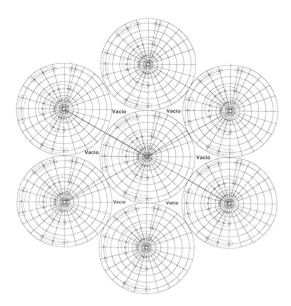

Figure 79 Set of 7 Tangent Miniverses with intergalactic movement 2D view

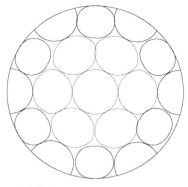

Figure 80 Set of 7 Tangent Miniverses with intergalactic motion 3D view

The Miniverses, as we had stated in chapter 4, are fully autonomous and function separately on the 56 axes of the universe. Figures 79 and 80 show empty spaces between the Miniverses. These empty spaces do not contain the fabric of space-time since it was formed only within the spherical caps of space-time, as we explain at length in the chapter 5

18 THE END OF THE UNIVERSE

The tendency of the Universe is to continue growing indefinitely in size, mass and energy. This increasing energy accumulates in equal parts in each of the Miniverses. Will there be a limit to the amount of energy each Miniverse can store?

But not all created matter in the universe is preserved forever. All systems of known nature have a recycling system, where old elements are replaced by new elements. The same is true in the universe, in whose recycling system black holes play a leading role.

Supermassive black holes, located at the center of galaxies, permanently engulf the oldest stars, storing them within their singulariy.

The recycling of matter by primordial black holes is of enormous importance because their function is not limited to the recycling of galaxies, since they are also the engine of the recycling of the universe itself.

As we demonstrate in chapter 15.2, primordial black holes, located in the Miniverses, have an irreversible tendency to increase their density and temperature towards infinite values, repeating in each of them the conditions that originated the Big-Bang, which created our universe.

18.1 From the Miniverses to the Fractal Multiverse

Our destiny is clearly defined, it is only a matter of time. At the inexorable moment, in the 56 miniverses there will be an explosion similar to the one that gave rise to our universe. That is 56 new Big-Bangs that will give way to 56 new universes. Each of these new universes will also end their lives generating another 56 new universes, and so on indefinitely in a geometric progression of order 56.

In the same way we conclude that our universe originated in one of the 56 Miniverses that exploded at the end of life of an earlier universe, which in turn had originated from the explosion of one of the 56 Miniverses of an even more earlier universe following in countdown the same geometric progression of order 56.

It's truly amazing, a fractal Multiverse up and down. An infinite chain reaction.

The Multiverse of which we are a part, is similar to a family tree of infinite single-parent families of 56 descendants.

Not even the sharpest minds of science fiction could have imagined such a dramatic, yet majestic and beautiful ending. The life of humanity is assured in time and space. It has always existed and will always exist, since the probability of its existence in an infinite universe in space and time is 100%.

In Figure 81, we can see part of our family of universes.

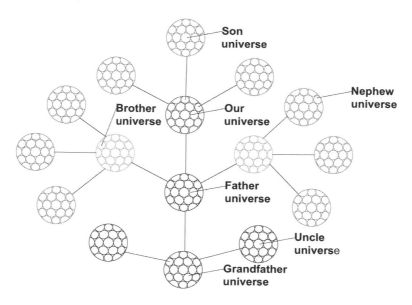

Figure 81 Fragment of the Multiverse

Part of our universal single-parent family of 56 descendants. In green two of our 55 Sister Universes, which currently coexist with our universe in blue. In magenta our future descendant universes: 3 of 56 Universes Sons and 6 of 3,080 Universes Nephews. In black part of our now extinct ascending universes: a Father Universe; two of our 55 Uncle Universes and a Grandpa Universe. The Multiverse houses a number of infinite families similar to ours.

It's an ideal ending that satisfies everyone. Gone is the disturbing concern about what was before this universe and the no less troubling one about what will come next. This ending is also compatible with almost all theories about the origin and destiny of the universe. It is valid for those who thought that the universe was generated with the Big Bang, but it is also compatible with those who assumed that the universe was infinite. It is also compatible with those who supported the theory of the Multiverses. Being valid for all, it is beautiful, it seems the work of a higher being. This extraordinary ending, by analogy, gives us the foundations of our last theory which completes the puzzle, joining the end with the beginning, the explanation of the origin of the Big Bang:

19 THE ORIGIN OF THE BIG-BANG

The Explosion called Big-Bang, which originated our universe, owes its origin to the explosion of a primordial black hole belonging to one of the 56 Miniverses of a universe prior to ours.

Bernard Carr and a group of physicists are in search of black holes older than the Big Bang.

Bernard Carr, emeritus professor of mathematics and astronomy at Queen Mary University, says in his article for New Scientist[31] that this is a far-fetched idea, but not inconceivable. *"The existence of primordial black holes formed in this universe is speculative, so the notion of black holes from an earlier universe might seem doubly speculative,"*

Carr said. *"However, it is important to explore this possibility, if not stimulating. Just as thinking about primordial black holes has led to important discoveries about quantum gravity, thinking about black holes before the Big Bang can lead to new physical discoveries, even if it turns out that the universe is not cyclical..."*

20 CONCLUSIONS

We believe that our Theory of the New Cosmological Model has fully fulfilled the function of scientifically explaining the origin, evolution and final destination of the universe, by providing us with a model that allows us to calculate its dimensions and analyze its behavior at any time. It thus clarifies, diaphanously, mysteries unexplained until now, such as dark energy, cosmic inflation and dark matter. It is discovered, among others, that gravity was created as a result of the expansion of the universe, that the energy of the universe comes from the Big Bang, which grows permanently by the perennial motion of its primordial particles as an inexhaustible source of energy. We have revealed, how galaxies and stars were created, the details of the components and how stellar black holes and primordial black holes operate and their leading role in the origin, functioning and fate of the universe.
It is worth noting the transcendental role that the principle of convertibility of energy in Matter and vice versa has in the formulation of this new vision of the universe, as established by Albert Einstein's Theory of

---

[31] Una nueva investigación sugiere que hay agujeros negros más antiguos que nuestro universo - Vandal Random (elespanol.com)

Special Relativity. Using the equation $E = \sqrt{p^2c^2 + m^2c^4}$ we were able to prove without any doubt:

1) The creation of the universe
2) The shape and dimensions of the universe
3) The expansion of the universe.
4) The creation of matter and space-time.
5) The creation of primitive stars and galaxies.
6) The movement and speed of the bodies in space-time
7) The curvature of space-time and gravity.
8) The shape and operation of black holes and their potential energy
9) The movement of stars within galaxies.
10) The intergalactic movement.
11) The energy that will create the new Big-Bangs and consequently, by analogy, the origin of the Big-Bang, and the conclusion that we are part of an infinite Fractal Multiverse in space and time
12) Undoubtedly we owe Albert Einstein the conception of the most important scientific idea in the history of science

$E = \sqrt{p^2c^2 + m^2c^4}$ it's an equation of the size of the universe.

But the most important thing about this research is that the model allows us to reveal what existed before our universe originated, as well as what will be after its existence has been exhausted since we undoubtedly belong to an infinite fractal Multiverse.

This revelation leads us to a final conclusion of great significance:

"**The universes and their laws of classical physics which include time and space-time are not permanent; they belong to a virtual reality. They're like soap bubbles that inflate and disappear. On the contrary, energy and matter are timeless and permanent, which implies the absence of time and space-time in the laws of Quantum Mechanics and explains incomprehensible phenomena for our virtual universe, such as 1) Quantum Leap and 2) Quantum Entanglement. The first shows that time does not exist in Quantum Mechanics and the second that space-time also does not exist in that science that studies of matter itself and its components**".

However, it is a theory, an approximation to reality, as is the conventional theory, of the Standard Cosmological Model, also based on the Big Bang. and it's the theory on which the study pensum and research programs of the science of astrophysics are structured. Which of these two approaches will be closer to reality? One way to elucidate this controversy is to check the quantity and quality of the evidence that underpins each theory.

Evidence can be of various types:

- Astronomical observations, confirming, in whole or in part, the theory under study.
- Mathematical demonstrations and other results, or products arising from theory that are coincident with previously proven realities.

We will first analyze the evidence of the Theory of the New Cosmological Model and then analyze the evidence of the Theory of the Standard Cosmological Model.

20.1 Analysis of the evidence of proof of the New Cosmological Model

For this analysis in the case of the New Cosmological Model, considering that it is a theoretical body of 40 theories, intertwined as a puzzle existing continuity between them; we can elaborate a flowchart that joins the beginning (Big Bang) with the end, the theory 40, which corresponds precisely the origin of the Big Bang.

TABLE 23 NEW COSMOLOGICAL MODEL FLOWCHART

THEORY	EVIDENCE: CONFIRMATION/DENIAL
T1 The Big-Bang Theory The mass of a singularity increases its density and temperature to almost infinite values and produces a large explosion	1--Hubble observations 2-Métric Friedman, Lemaitre, Robertson, Walker 3-CMB discovery
T2 Origin theory of the universe The expansive force of the Big Bang produces a chain of nuclear fissions that reduce the density of the primordial particles to the lighter element, hydrogen. They are "n" particles that are e awarded in all directions with divergent angle "Θ" at the speed of light	Equation of the Theory of Special Relativity: $$E = \sqrt{p^2c^2 + m_0^2c^4}$$ Explains the conversion of a maximum density rest mass to "n" minimum density masses moving at the maximum speed of light.

T3 Theory Form and dimensions of the universe The universe is spherical with a radius equal to 1.9318 rO (rO=radius observable universe). With an internal CMB sphere of radius 0.9318 rO	Astronomical observations of the observable universe allow geometrically to size the universe and calculate the angle "Θ"=30° and the number of primeval particles "n"=56 Research published by the journal Nature Astronomy [32] reveals the existence of evidence in the CMB, leading to a spherical closed universe, which coincides with our theory
T4 Teoría Los Miniversos The divergent trajectory of the 56 primordial particles at the speed of light causes the universe to evolve separately in 56 Miniverse homogeneously in all directions	Astronomical observations indicate that the universe is homogeneous and isothermal in all directions
T5 Cosmic inflation theory never occurred	The theory of the Miniverses makes the theory of Cosmic Inflation unnecessary
T6 Teoría: La creación de la materia y el espacio-tiempo Due to the momentun produced by its displacement at the speed of light, its mass doubles, the mass created as clones is deposited on the adjacent axes. Creating simultaneously matter and the space-time where it is deposited	$E = \sqrt{m_p^2 c^4 + m_p^2 c^4}$ Einstein's equation proposes a vector sum and since the velocity of both terms has equal direction and meaning, it corresponds to the sum of two parallel vectors. Therefore we add algébricamente its modules: $E = m_p c^2 + m_p c^2 = 2\, m_p c^2$ **$E = 2\, m_p c^2$** Therefore a body that moves at the speed of light will double its energy and consequently its mass creating a succession of clones of the primordial matter.
T7 Theory Cause of the expansion of the universe	The energy equations: 1. $E = \sqrt{p^2 c^2 + m_0^2 c^4}$

---

[32] Planck evidence for a closed Universe and a possible crisis for cosmology | Nature Astronomy

The universe is continuously expanding at the speed of light due to the growth of matter and the space where it is housed as a result of the energy generated by the perpetual motion, at the speed of light, of its 56 primeval particles now transformed into miniverses.	2. $m = \dfrac{m_o}{\sqrt{1-\dfrac{v^2}{c^2}}}$ 3. $x = (dx + c\,t)$; $y = (dy + c\,t)$; $z = (dz + c\,t)$; $Z = (dZ + ct)$ They show that matter and space-time, where it is housed, grow and move permanently at the speed of light.
T8 Dark energy is unnecessary in the New Cosmological Model	The previous theory T7 of the cause of the expansion of the universe makes the theory of dark energy unnecessary
T9: The creation of the first stars and galaxies in the universe The formation of the multiple galaxies occurred at the beginning of the universe	1) Late anisotropy confirms this theory 2) Recent observations[33] made using the ALMA telescope show the appearance of galaxies in the early universe 3) Sharp images from the James Web Supertelescope (JWST) reveal the existence of galaxies in the universe even earlier, just 320 million years after the Big Bang [34].
T10: The Special Theory of Relativity is modified by eliminating the concept of time dilation with speed	It is mathematically demonstrated in the development of this theory that in the expanding universe time does not dilate with speed. What happens is that the trajectory grows
T11: Scientific explanation of Hubble's law	It is mathematically proven that the expansion model of the New Cosmological Model complies with Hubble's Law and explains the scientific cause of it

---

[33] Detectan colosal tormenta de agujero negro en el universo primitivo, la más antigua hasta la fecha | Ciencia y Ecología | DW | 18.06.2021
[34] Hito histórico: localizaron la galaxia más lejana en el Universo (clarin.com)

T12: Current age of the universe The current age of the universe is $13,978 \times 10^9$ years	With the data of the Hubble constant, $H0 = 70 Km/sec/Mpc$ provided by the WMAP satellite, applied to the NCM expansion model, verified by hubble law, the current age of the universe is calculated.
T13: The spherical cap of the curvature of the surface of space-time;. T14: The cause of gravity and the meaning of the gravitational constant G and its exact theoretical value; T24 Theory Modification of the law of universal gravitation	The exact calculation of the gravitational constant $G=(\cos 0.5\ radian)^3$ performed in the development of the T14, demonstrates the validity of these three theories
T15 Spheroidal gravitational fields, T16 The cause of the inclination of the axis of rotation of the Earth and other planets, T17 the planetary orbits according to Kepler's 1st law	These three theories are confirmed with the scientific proof of Kepler's law and the experiment of the almost exact calculation of Saturn's elliptical orbit only with the data of the eccentricity of the planet and its obliquity
T18 Time is not modified by gravity, T19 Einstein's Theory of General Relativity is ruled out because of its incompatibility with the expanding universe	Both theories are discarded because of their incompatibility with the expansion of the universe and because the model of curvature of space-time used by Einstein is contradictory to Kepler's 1st law
T20 Light is observed by describing a curved trajectory in the universe	Eddington's experiment in 1919 in observing an eclipse erroneously attributed to Einstein's General Theory of Relativity is evidence of this theory
T21 The angular acceleration provided for in Kepler's 2nd Law resolves the apparent anomaly of Mercury's orbit	The precise calculation of the angular acceleration due to the great eccentricity of mercury's orbit clarifies that there is no anomaly in its orbit
T22 The rotating elliptical orbits of the Sun: The sun rotates an orbit through each planet similar to that of the planet around the Sun	It is shown that the kinetic energy of the Sun traveling through its orbit is exactly equal to the kinetic energy of the planet traveling through its orbit and is also verified with astronomical observations, of the precession of the orbit of Mercury.

T23 Modification of Kepler's laws: It should be noted that in each ecliptic plane there are two orbits one of the planet and a similar one traveled by the Sun.	The evidences of theory 22, lead to the formulation of theory 23
T24 Modification of the Law of Universal Gravitation	The exact calculation of the gravitational constant $G(\cos 0.5 \text{ radian})^3$ made in the development of Chapter 10 demonstrates the validity of this amendment to the Law of Universal Gravitation
T25 Operation and dimensions of a stellar black hole	The calculation of the dimensions of the supermassive black hole of the galaxy NGC 4258 is a test evidence of this theory.
T26 The drag flow of space-time at constant speed into the interior of a rotating black hole, called kerr flow	The observations and measurements made by Vera Rubin and Kent Ford on the constant velocity of stars on the periphery of galaxies is evidence of this theory.
T27. In the Gravitational Zone of Influence 10% of the distance to the center of the supermassive black hole of galaxies, the trajectory of the particles towards the black hole, follow the path of the Kerr flow and the speed according to the Law of Universal Gravitation	Observations and measurements of the motion of the masers in the accretion disk of the galaxy NGC 4258, confirms this theory.
T28 The Space-time Drag Force of a Stellar Black Hole	The numerous observations of the merger of two black holes are evidence of this theory
T29 Tidal disruption event	The numerous astronomical observations of these events confirm this theory.
T30 The Dark Theory of Dark Matter	The formulation of the Dark Matter Theory contrary to the scientific method and the incorrect use of the law of universal gravitation in its formulation clearly rule out its validity

T31 Origin of the galactic bulb	Recent research[35] by an international scientific team led by Center for Astrobiology confirms this theory
T32 Origin of the accumulation of stars in the nucleus of galaxies..	Astronomical observations confirm the elliptical shapes of the illuminated part of the galactic nucleus calculated in this theory
T33: Funcionamiento and dimensions of the components of a primordial black hole	There is no evidence for this theory
T34 The drag flow of space-time into the interior of a primordial black hole	The Dark Flow and the Great Attractor are evidence of this theory.
T35: The Cause of the Dark Flow	Theory 34 provides a theoretical explanation of Dark Flow
T36 The gravitational energy of a primordial black hole	It is demonstrated based on equations (E1-1), (E1-2) and (E1-3)
T37 origin of the Quasars	$$F_\mathrm{p} = 0,255 \frac{R}{rO} C \quad (E\,12-6)$$ The equation (E12-6), reproduces the astronomical observations, which reflect that the luminosity and dimensions of quasars increase with their age
T38 Primordial flux determines the motion of galaxies	The exact reproduction of the behavior of galactic Groups, Clusters and Superclusters in an experiment using chapter 17 is a verifiable evidence of the same.
T39 From the Miniverses to the Fractal Multiverse	The equation (E15-10), called the "Big Bang Equation", shows mathematically that the singularities of the primordial black holes of the 56 miniverses irreversively brought their density and temperature to almost infinite values, thus repeating

---

[35] Identifican el origen de estructuras formadas en galaxias como la Vía Láctea - Ciencia - ABC Color

	the conditions that originated the Big Bang
T40 The origin of the Big Bang	It follows by analogy

From the analysis of this flowchart we conclude:

"The Theory of the New Cosmological Model, presents a description, total and complete of the origin, evolution and final destination of our universe, based on a set of scientific theories, corroborated by multiple confirmatory events".

20.2 Analysis of the evidence of the Standard Cosmological Model Theory (ECM)

The Standard Cosmological Model was not developed as a whole, so there is no binding relationship between its theories. However we can create a flowchart by sorting their main theories from the origin of the universe and ending with theories related to their fate. In Table 24 we can see the flowchart of the Standard Cosmological Model.

Table 24 STANDARD COSMOLOGICAL MODEL FLOWCHART

THEORY	EVIDENCE: CONFIRMATION/DENIAL
Big-Bang Theory The mass of a singularity increases its density and temperature to almost infinite values and produces a large explosion	1- Hubble Obsevations. 2- Friedman, Lemaitre, Robertson, Walker Metric. 3- CMB discovery
Theory of the creation of primal matter Due to the energy of the Big Bang explosion, the mass of singularity disintegrated completely even at the molecular level, disappearing the atoms and their components. For this reason there was no remaining energy from that of the explosion. Subsequently the Standard Cosmological Model states that	According to this theory, the chain of nuclear fissions did not stop when it reached the hydrogen and continued with a process unknown by physics such as the total destruction of matter, then giving way to a creationist process. Two difficult-to-answer questions arise Is the total destruction of matter possible?, Is it possible to create matter from sub-atomic particles?

matter is reborn from the most elementary subatomic particles, creating the first atoms, then the first hydrogen molecules and after helium	Due to total disintegration there is no remaining energy from the Big-Bang and consequently there was no energy left for the development and expansion of the universe. In addition, a universe whose primary matter is static does not create time for its evolution and development. Since one of the consequences of movement is the creation of time. If there is no movement, time does not exist. It is the scene of a universe failed for lack of energy and the absence of time for its evolution
Cosmic Inflation Theory The pre-primal universe, expanded exponentially for a short time. To explain in this way that it is homogeneous, isotropic and isothermal in all directions.	This theory arises as a result of the deficiencies of the theory on the creation of primal matter according to the Standard Cosmological Model. The primal matter according to that model was static and without energy to move. In that case when the universe expands from a single point it cannot be explained that it is homogeneous, isotropic and isothermal Cosmic inflation solves this problem but creates a more complicated one to solve where energy comes from for this exponential expansion of primeval matter.
Dark Energy Theory. The universe expands accelerated due to unknown energy. Which makes up 70% of the energy mass of the universe.	This theory arises as a result of the deficiencies of the theory on the creation of primal matter according to the Standard Cosmological Model. The primal matter in

Dark energy is invisible and its provenance is not known. It is assumed to exist simply because the primal matter of the Standard Cosmological Model lacks energy for the expansion of the universe.	that model is static and without the energy needed for the expansion of the universe..
Theory of Dark Matter The theory states that there is an invisible matter more abundant than visible matter and represents 25% of the matter-energy of the universe Its existence is inferred from the non-compliance with the law of universal gravitation in measuring the speeds of the stars of the periphery of galaxies	The formulation of the Theory of Dark Matter contrary to the scientific method and the incorrect use of the law of universal gravitation in its formulation clearly rule out its validity.
Gravity theorians at the ECM The Standard Cosmological Model uses three antagonistic gravity theories between which we will analyze separately. 1) Gravity according to Newton's theory. 2) Gravity according to Einstein's theory of relativity. 3) Gravity as a primary fundamental force of nature.	Neither of the three versions explains the scientific reason for its existence or what does the gravitational constant G used by the first two versions
Newton's Theory of Gravity It's Isaac Newton's theory in the Law of Universal Gravitation It is used for the study of planetary systems, galaxies and the intergalactic system.	1) It does not define the exact meaning and theoretical value of the gravitational constant G nor the origin of gravitational energy. 2) It is inappropriate to use it to explain the movement of stars within galaxies, so

	to try to cover their deficiency, dark matter is used. 3) It is inappropriate to use it to explain the intergalactic movement, so to try to cover its deficiency, dark matter is used.
Theory of Gravity According to Einstein's Theory of General Relativity According to this theory gravity is not a force, what exists is a geometric phenomenon due to the deformation of space time proportional to the mass of bodies. Used to analyze stellar black holes.	1) It does not define the exact meaning and theoretical value of the gravitational constant G, nor the provenance of gravitational energy. 2) The shape of the curvature of space time whose dimensions are proportional to the mass of the bodies makes it incompatible with the 1st law of Kepler. Which we were able to demonstrate in chapter 10. 3) The erroneous explanation of Mercury's alleged orbit anomaly disqualifies it
Theory of Gravity as Nature's Primary Fundamental Force. According to this theory gravity is an extracorporeal primary force that exists independent of the masses of bodies and that prevents them from separating by the expansion of the universe. Contrary to Newton's law which states that the gravitational pull between two bodies creates two equal and opposite-sense forces proportional to their masses. In this case it is a force that acts	1) It does not use the gravitational constant G and does not explain the provenance of its energy. 2) Does not comply with Kepler's laws. 3) Its operation is incompatible in conjunction with the other two versions of gravity.

on all bodies at once. contrary to their trajectory under Hubble's law and all are headed to the same focal point located at the origin of the Big Bang. It is used to explain the theory called Big Crunch	
Theory of Conflict Between Dark Energy vs. Gravity and Dark Matter According to this theory, gravity reinforced by dark matter opposes the expansion of the universe driven by the Dark Energy, which poses an endless struggle and the result of which will eventually define the fate of the universe. With two possible outcomes: 1) Gravity and Dark Matter overcomes. In this case the fate of the universe will be according to the Big Crunch. 2) Overcome the Dark Energy. In this case the fate of the universe will be according to the Big Frezee theory	The antagonistic struggle between gravity and the expansion of the universe is totally wrong since gravity is the product of the expansion of the universe, as demonstrated in chapter 10.1 of this investigative work. The opposite is true if the universe did not expand, there would be no gravity
Big Crunch Theory If gravity is imposed the universe contracts itself, in the so-called Big Crunch, a kind of Big Bang but on the contrary.	It is discarded for the reasons set out in the above theory and further than overcoming the expansion force produced by the Big Bang would require a higher antagonistic energy source than the Big Bang

Big Freeze Theory	
If the Dark Energy were imposed, the end of the universe would be the so-called Big Freeze or thermal death of the universe. The universe will continue to expand as before, now, but the stars will separate and shut down and the universe will turn dark and cold.	Big Freeze, is an unworkable destination for the universe, for if the universe expanded eternally, it would look as it is today and as it has always been. It should be remembered that the metric expansion of the universe, establishes that with the expansion grows at the speed of light space but also the mass of bodies and energy, so the thermal death of the universe is not possible

On this theory we have the following opinion:

The theory of the Standard Cosmological Model on the origin, evolution and destiny of the universe, in our opinion, is unworkable and completely wrong, because of the multiple negative evidences of the theories that underpin it:

After the analysis of both theories and the quality and quantity of their evidence, our conclusion is obvious in favor of the Theory of the New Cosmological Model. This encourages us to continue researching this new path and try to understand our immense and incredible universe.

But it's just our opinion, which is irrelevant.

They have a responsibility to take the floor and give an opinion the universities, academies of sciences, and research institutions of the science of astrophysics, which are responsible for the teaching and research of this science

21 ACKNOWLEDGEMENTS

My thanks to the engineers Cristóbal Enrique Lander Rodríguez, Carlos Alberto Lugo Piñerua and Edgard Humberto Paredes Pisani y Antonio Coso for their collaboration in the development of this research. Also my thanks to Wikipedia, source of permanently updated scientific data.

22 REFERENCES

1. Planck evidence for a closed Universe and a possible crisis for cosmology | Nature Astronomy
2. ¿Vivimos en un Universo en forma de bucle? (abc.es)
3. https://es.wikipedia.org/wiki/Inflaci%C3%B3n c%C3%B3smica
4. https://es.wikipedia.org/wiki/Expansi%C3%B3n m%C3%A9trica del espacio#:~:text=La%20expansi%C3%B3n%20m%C3%A9trica%20del%20espacio,hace%20m%C3%A1s%20joven%20o%20viejo
5. https://es.wikipedia.org/wiki/Energ%C3%ADa oscura
6. https://en.wikipedia.org/wiki/Cosmic microwave backgroun
7. Detectan colosal tormenta de agujero negro en el universo primitivo, la más antigua hasta la fecha | Ciencia y Ecología | DW | 18.06.2021
8. Hito histórico: localizaron la galaxia más lejana en el Universo (clarin.com)
9. https://en.wikipedia.org/wiki/Hubble's law
10. https://es.wikipedia.org/wiki/Esferoide
11. https://es.wikipedia.org/wiki/Geometr%C3%ADa descriptiva
12. Black hole - Wikipedia
13. Stellar black hole - Wikipedia
14. Intermediate-mass black hole - Wikipedia
15. Supermassive black hole - Wikipedia
16. Primordial black hole - Wikipedia
17. https://es.wikipedia.org/wiki/Horizonte de sucesos
18. https://es.wikipedia.org/wiki/Ergosfera
19. https://en.wikipedia.org/wiki/Accretion disk
20. **"Alberdi Odriozola, Antxon. Los agujeros negros (NATGEO CIENCIAS) (Spanish Edition) . National Geographic. Edición de Kindle**
21. Identifican el origen de estructuras formadas en galaxias como la Vía Láctea - Ciencia - ABC Color
22. Bulbo galáctico - Wikipedia, la enciclopedia libre
23. La materia oscura está ralentizando la barra de la Vía Láctea (europapress.es)
24. https://www.youtube.com/watch?v=qKLQbyr0Clk
25. Primordial black hole - Wikipedia
26. Researchers decipher the history of supermassive black holes in the early universe - Media Relations (uwo.ca)
27. https://www.space.com/5878-mysterious-dark-flow-discovered-space.htm
28. https://es.wikipedia.org/wiki/Flujo oscuro#:~:text=En%20cosmolog%C3%ADa%20f%C3%ADsica%2C%20el%20flujo,constelaciones%20de%20Vela%20y%20Centauro
29. https://en.wikipedia.org/wiki/Quasar
30. *https://es.wikipedia.org/wiki/Agrupaciones galácticas*

23 AUTHOR DATA

The author (Spanish-Venezuelan) is an Electrician Engineer graduated from the Central University of Venezuela, with postgraduate studies in Business Administration.

He has extensive experience of more than 30 years in research, design and calculation of multidisciplinary engineering projects (electrical, civil and mechanical) in different sectors, as President and founder of ERIPE, CONSULTING ENGINERS, C.A. in Venezuela and Colombia.

Since 2015 dedicated to independent research in astrophysics. Has published the following books of astrophysics: Integral Theory of the Universe, Amazon 2020; The origin of the gravity and the value exact Constant G, Amazon, 2019; The shape and dimension of the universe, Amazon, 2019; Revealed the mysteries of the dark matter dark energy and Cosmic inflation, Amazon, 2019
Born in Barcelona (Venezuela), he has lived since 2009 in Alicante (Spain).

Enrique Lander R.
enrique.lander@gmail.com
http://amazon.com/author/enrique

Printed in Great Britain
by Amazon

34129788R00105